James J. Buckley, Leonard J. Jowers

Simulating Continuous Fuzzy Systems

Studies in Fuzziness and Soft Computing, Volume 188

Editor-in-chief
Prof. Janusz Kacprzyk
Systems Research Institute
Polish Academy of Sciences
ul. Newelska 6
01-447 Warsaw
Poland
E-mail: kacprzyk@ibspan.waw.pl

Further volumes of this series can be found on our homepage: springeronline.com

Vol. 174. Mircea Negoita, Daniel Neagu, Vasile Palade
Computational Intelligence: Engineering of Hybrid Systems, 2005
ISBN 3-540-23219-2

Vol. 175. Anna Maria Gil-Lafuente
Fuzzy Logic in Financial Analysis, 2005
ISBN 3-540-23213-3

Vol. 176. Udo Seiffert, Lakhmi C. Jain, Patric Schweizer (Eds.)
Bioinformatics Using Computational Intelligence Paradigms, 2005
ISBN 3-540-22901-9

Vol. 177. Lipo Wang (Ed.)
Support Vector Machines: Theory and Applications, 2005
ISBN 3-540-24388-7

Vol. 178. Claude Ghaoui, Mitu Jain, Vivek Bannore, Lakhmi C. Jain (Eds.)
Knowledge-Based Virtual Education, 2005
ISBN 3-540-25045-X

Vol. 179. Mircea Negoita, Bernd Reusch (Eds.)
Real World Applications of Computational Intelligence, 2005
ISBN 3-540-25006-9

Vol. 180. Wesley Chu, Tsau Young Lin (Eds.)
Foundations and Advances in Data Mining, 2005
ISBN 3-540-25057-3

Vol. 181. Nadia Nedjah, Luiza de Macedo Mourelle
Fuzzy Systems Engineering, 2005
ISBN 3-540-25322-X

Vol. 182. John N. Mordeson, Kiran R. Bhutani, Azriel Rosenfeld
Fuzzy Group Theory, 2005
ISBN 3-540-25072-7

Vol. 183. Larry Bull, Tim Kovacs (Eds.)
Foundations of Learning Classifier Systems, 2005
ISBN 3-540-25073-5

Vol. 184. Barry G. Silverman, Ashlesha Jain, Ajita Ichalkaranje, Lakhmi C. Jain (Eds.)
Intelligent Paradigms for Healthcare Enterprises, 2005
ISBN 3-540-22903-5

Vol. 185. Dr. Spiros Sirmakessis (Ed.)
Knowledge Mining, 2005
ISBN 3-540-25070-0

Vol. 186. Radim Bělohlávek, Vilém Vychodil
Fuzzy Equational Logic, 2005
ISBN 3-540-26254-7

Vol. 187. Zhong Li, Wolfgang A. Halang, Guanrong Chen
Integration of Fuzzy Logic and Chaos Theory, 2006
ISBN 3-540-26899-5

Vol. 188. James J. Buckley, Leonard J. Jowers
Simulating Continuous Fuzzy Systems, 2006
ISBN 3-540-28455-9

James J. Buckley
Leonard J. Jowers

Simulating Continuous Fuzzy Systems

Springer

Professor James J. Buckley
Department of Mathematics
University of Alabama at Birmingham
35294-1170 Birmingham
U.S.A.
E-mail: buckley@math.uab.edu

Leonard J. Jowers
Department of Computer and
Information Sciences
University of Alabama at Birmingham
35294 Birmingham, Alabama
U.S.A.

ISSN print edition: 1434-9922
ISSN electronic edition: 1860-0808
ISBN-13 978-3-642-06668-9
e-ISBN-13 978-3-540-31227-7

Springer is a part of Springer Science+Business Media
springeronline.com

Softcover reprint of the hardcover 1st edition 2006

To Julianne and Helen,

To Paula and “the kids”.

Contents

Chapter 1

Introduction

1.1 Introduction

This book is written in two major parts. The first part includes the introductory chapters consisting of Chapters 1 through 6. In part two, Chapters 7-26, we present the applications. This book continues our research into simulating fuzzy systems. We started with investigating simulating discrete event fuzzy systems ([7],[13],[14]). These systems can usually be described as queuing networks. Items (transactions) arrive at various points in the system and go into a queue waiting for service. The service stations, preceded by a queue, are connected forming a network of queues and service, until the transaction finally exits the system. Examples considered included machine shops, emergency rooms, project networks, bus routes, etc. Analysis of all of these systems depends on parameters like arrival rates and service rates. These parameters are usually estimated from historical data. These estimators are generally point estimators. The point estimators are put into the model to compute system descriptors like mean time an item spends in the system, or the expected number of transactions leaving the system per unit time. We argued that these point estimators contain uncertainty not shown in the calculations. Our estimators of these parameters become fuzzy numbers, constructed by placing a set of confidence intervals one on top of another. Using fuzzy number parameters in the model makes it into a fuzzy system. The system descriptors we want (time in system, number leaving per unit time) will be fuzzy numbers. In general, computing these fuzzy numbers can be difficult. We showed how crisp discrete event simulation can be used to estimate the fuzzy numbers used to describe system behavior.

This book is about simulating continuous fuzzy systems. Continuous systems, or continuous time dynamical systems, are usually described by a system of ordinary differential equations (ODEs). Many parameters in the system of ODEs are not known precisely and must be estimated. To show the uncertainty in these parameter values we will use fuzzy number estimators. Fuzzy number parameter values produce a system of fuzzy ODEs to solve and we have a continuous fuzzy system, or a continuous time fuzzy dynamical system. Solution trajectories become fuzzy trajectories. A cut through the fuzzy trajectory at any time t produces a fuzzy number. We plan to use crisp continuous simulation to estimate these fuzzy trajectories.

But first we need to be familiar with fuzzy sets. All you need to know about fuzzy sets for this book comprises Chapter 2. For a beginning introduction to fuzzy sets and fuzzy logic see [9].

Chapter 3 gives a brief introduction to fuzzy estimation. We will use only two methods of fuzzy estimation: from expert opinion or from data. We explain how you can get fuzzy numbers when you estimate, from crisp data, probabilities or parameters in probability densities. The basic construction involves placing confidence intervals, one on top of another, to obtain a fuzzy number as our estimator instead of using a point estimator or a single confidence interval.

Chapter 4 introduces continuous fuzzy (dynamical) systems theory. Consider a system of differential equations whose solution describes the evolution of the crisp continuous (dynamical) system. This system of differential equations usually has a number of parameters many of whom their values are not known precisely. To show the uncertainty in these parameter values we will use fuzzy number estimators (Chapter 3). Fuzzy number parameter values produce a system of fuzzy ODEs to solve and we have a continuous (dynamical) fuzzy system. We plan to use crisp continuous simulation to estimate the base of these fuzzy trajectories, which we will call the band of uncertainty. In the rest of this book we will call a crisp continuous time dynamical system simply a continuous system and a continuous time fuzzy dynamical system simply a continuous fuzzy system.

How do we choose simulation software to accomplish all the simulations in Chapters 7-26 is the topic of Chapter 5. We discuss cost, ease of use, need to run on a desktop computer, plus some other concerns we consider in selecting the simulation software. Our final decision is also discussed.

Chapter 6 introduces a type of simulation optimization. We discuss how we plan to solve the simulation optimization problems presented in Chapters 7-26. The general problem remains unsolved. Let us briefly discuss this optimization problem. Some parameter values are uncertain and we use their fuzzy number estimators. Let these parameters range throughout the interval which is the base of the fuzzy number. We obtain an infinite number of crisp solutions (trajectories) which we call the uncertainty band. We want to determine and graph the boundary of this uncertainty band. This is the topic of Chapter 6 .

The structure of the rest of the book is now determined. Use continuous simulation to approximate the boundary of the uncertainty bands for the fuzzy systems. The crisp system is usually sufficiently complicated so that the exact crisp solution is either too difficult to work with (to correctly fuzzify), or we do not have an exact closed-form mathematical solution. We need to use software to obtain graphs of the solutions. This will be the topic of Chapters 7-26.

The applications in Chapters 7-26 are quite varied ranging from predator/prey models to bungee jumping to a human cannon ball showing the varieties of continuous fuzzy systems. These chapters may be read independently. This means some material, including a discussion of the system of differential equations, fuzzy estimators for some of the parameters producing a fuzzy system, the optimization problem, the simulation diagram, etc., is repeated in each chapter.

How we organized the continuous simulation program is shown in each Chapter 7 - 26. We did not need to write any computer code to use our simulation software if we wanted to obtain only one solution graph per simulation run. Simulation operations are represented as icons and we connect them with arrows using the mouse. Diagrams showing the icons and connecting arrows are given for each application in Chapters 7 - 26. However, we wanted to place up to 729 solution graphs in a single figure; hence, we had to write Matlab code in order to get this result. Further details are in Chapter 28.

This book is based on, but expanded from, the following recent papers and publications: (1) fuzzy estimation, probability and statistics ([4]-[6],[12]); (2)fuzzy systems [8]; and (3) simulating continuous fuzzy systems ([10],[11]).

There are no prerequisites, but it would be helpful to know some basic information about ordinary differential equations (see Section 2.5). However, the reader should be able to understand, from the figures and analytical development, how the continuous simulation is useful in analyzing continuous fuzzy systems.

1.2 Notation

It is difficult, in a book with a lot of mathematics, to achieve a uniform notation without having to introduce many new specialized symbols. Our basic notation is presented in Chapter 2. What we have done is to have a uniform notation within each chapter. What this means is that we may use the letters "a" and "b" to represent a closed interval $[a, b]$ in one chapter but they could stand for parameters in a differential equation in another chapter. We will have the following uniform notation throughout the book: (1) we place a "bar" over a letter to denote a fuzzy set ($\overline{A}$, $\overline{B}$, etc.), and all our fuzzy sets will be fuzzy subsets of the real numbers; and (2) an alpha-cut of a fuzzy set (Chapter 2) is always denoted by "α". Since we will be using α for alpha-cuts we need to change some standard notation in statistics: we use β in confidence intervals. So a $(1-\beta)100\%$ confidence interval means a 95% confidence interval if $\beta = 0.05$. When a confidence interval switches to being an alpha-cut of a fuzzy number (see Chapter 3), we switch from β to

α. All fuzzy arithmetic is performed using the extension principle (Chapter 2). The term "crisp" means not fuzzy. A crisp set is a regular set and a crisp number is a real number. Also, throughout the book $\overline{x}$ will be the mean of a random sample, not a fuzzy set.

1.3 Applications

All the applications, except Chapter 9, deal with systems (from two to five differential equations). Chapter 9 uses only one nonlinear ODE. About half of the systems are linear and the rest are nonlinear. The parameters in the systems, whose values may be uncertain, range from a minimum of two to a maximum of eleven.

We usually do not present complete derivations of the systems of ODEs. This is not a book on math modeling. Many times a complete derivation involves details from chemistry, biology, aeronautics etc. which is beyond the topic of this book. This is common practice in books on nonlinear ODEs where they present the system of ODEs and refer the reader to the original papers for the derivations.

In a number of applications the variables $x(t)$ $(y(t),z(t))$ represent the size of some population. Technically, $x(t)$ $(y(t),z(t))$ should then take on only positive integer values. However, it is common practice to model such systems using systems of ODEs and continuous variables so that $x(t)$ $(y(t),z(t))$ can take on any positive real number values. If we were to restrict $x(t)$ $(y(t),z(t))$ to be integer valued it may be better to work with systems of difference equations. We will not consider difference equations in this book, only differential equations.

1.4 Previous Research

Our approach to handling uncertainty in continuous systems in this book is not completely new. Methods of analyzing uncertainty in crisp differential equations has been going on for about twenty years. See ([1]-[3],[15],[17],[18]) for a review of this area. In this research the authors allowed uncertainty in the initial conditions, in the parameters in the differential equations and in some of the functional relationships between the variables in the equations. The uncertainty in the initial conditions and in the parameters is usually modeled using intervals but some authors employed fuzzy numbers. The analysis having fuzzy numbers always turns into using intervals after taking $\alpha-$cuts (Chapter 2). The uncertainty in functions is modeled by assuming their graphs lie between a pair of envelopes (an upper and lower graph). We will not assume any uncertainty in the structure of the differential equations, and functions, in our models.

The methods used in the study of uncertainty in crisp differential equations usually falls into two areas. They are the so called "AI-based methods", also called "semiquantitative simulation", and the Monte Carlo methods. In the semiquantitative method the object is to give a qualitative description of the behavior of all possible solutions. In the Monte Carlo technique the goal is to construct all possible solutions. However, the set of all possible solutions is infinite, so they compute some finite (discrete) approximation to the set of all possible solutions. Both areas have their advantages and disadvantages [17]. What we do in this book can be classified as the Monte Carlo method.

What then is new in this book is: (1) we argue in Chapter 4 that many crisp continuous (dynamical) systems naturally become fuzzy through fuzzy estimation (Chapter 3) of the uncertain initial conditions and parameters; (2) we find an approximation to the band of uncertainty which is the trajectory of the bases of the fuzzy number trajectories; and (3) we apply this to numerous diverse applications in Chapters 7 - 26 using the readily available simulation language Simulink [19].

1.5 Figures

The reader can see that there are three types of figures in this book. We now explain why we used three different types of figures.

1.5.1 Maple

Some of the figures, graphs of certain fuzzy numbers, in the book are difficult to obtain by standard methods (LaTeX) so they were created using a different method. These graphs were done first in Maple [16] and then exported to $LaTeX2_\epsilon$. We did these figures first in Maple because of the "implicitplot" command in Maple. Let us explain why this command was important in this book. Suppose $\overline{X}$ is a fuzzy estimator we want to graph. Usually in this book we determine $\overline{X}$ by first calculating its α-cuts. Let $\overline{X}[\alpha] = [x_1(\alpha), x_2(\alpha)]$. So we get $x = x_1(\alpha)$ describing the left side of the triangular shaped fuzzy number $\overline{X}$ and $x = x_2(\alpha)$ describes the right side. On a graph we would have the x-axis horizontal and the y-axis vertical. α is on the y-axis between zero and one. Substituting y for α we need to graph $x = x_i(y)$, for $i = 1, 2$. But this is backwards, we usually have y a function of x. The "implicitplot" command allows us to do the correct graph with x a function of y when we have $x = x_i(y)$. All figures in Chapters 2 and 3 were done in Maple and then exported to $LaTeX2_\epsilon$.

1.5.2 LaTeX

Some other figures were easily constructed using the graphics in $LaTeX2_\epsilon$. These figures are Figures 11.1, 12.1, 13.1-13.3, 15.1, 17.1, 22.1 and 28.1.

Color/Line Width	Description, graph generated by using only ...
red/3	left-supports, $\alpha = 0$ cut (Chapter 2)
black/2	cores, $\alpha = 1$ cut (Chapter 2)
blue/2	right-supports, $\alpha = 0$ cut (Chapter 2)
green/1	all others

Table 1.1: Color/Line Width Legend

1.5.3 Simulink

All other figures were constructed by Matlab/Simulink [19]. However, there was a problem of getting all the graphs for one variable on one coordinate system. In the system of differential equations describing the continuous system certain parameters were allowed to range throughout intervals. Using a finite choice of values for each uncertain parameter gave us at most 729 graphs to place on the same coordinate system. Special code to accomplish this is discussed in Chapter 28 since this is not automatic in this simulation software.

1.5.4 Color

We are able to use more than just black and white in the "on-line" publication of this book. Therefore, in many figures made from Matlab/Simulink we employed green, red, black and blue. Table 1.1 is the legend for the graphs. In the hard cover printing of the book the green will show up as grey. The black and green curves are plotted as unbroken lines. Blue curves are dot-dash lines. Red curves are dash-dash lines.

Access to the "on-line" publication is easily accomplished by surfing to `http://www.springerlink.com`. Find the "search for" box. Search for "Studies in Fuzziness and Soft Computing". On that page you will find a link to an on-line color version. Note however, that if you own the book and have access to Matlab/Simulink, you may prefer the option offered in Chapter 28 to explore the simulations more closely.

1.6 References

1. J.Armengol, L.Trave-Massuyes, J.Lluis de la Rosa and J.Vehi: Envelope Generation for Interval Systems, Proceedings CAEPIA 1997, Malaga, Spain, 33-48.

2. J.Armengol, L.Trave-Massuyes, J.Lluis de la Rosa and J.Vehi: On Modal Interval Analysis for Envelope Determination within the Ca-En Qualitative Simulator, Proceedings IPUM 1998, Paris, France, 110-117.

3. A.Bonarini and G.Bontempi: A Qualitative Simulation Approach for Fuzzy Dynamical Models, ACM Trans. Modeling and Computer Simulation, 4(1994)285-313.

4. J.J.Buckley: Fuzzy Probabilities: New Approach and Applications, Physica-Verlag, Heidelberg, Germany, 2003.

5. J.J.Buckley: Fuzzy Probabilities and Fuzzy Sets for Web Planning, Springer, Heidelberg, Germany, 2004.

6. J.J.Buckley: Fuzzy Statistics, Springer, Heidelberg, Germany, 2004.

7. J.J.Buckley: Simulating Fuzzy Systems, Springer, Heidelberg, Germany, 2005.

8. J.J.Buckley: Fuzzy Systems, Soft Computing. To appear.

9. J.J.Buckley and E.Eslami: An Introduction to Fuzzy Logic and Fuzzy Sets, Physica-Verlag, Heidelberg, Germany, 2002.

10. J.J.Buckley, K.Reilly and L.Jowers: Simulating Continuous Fuzzy Systems I, Iranian J. Fuzzy Systems. To appear.

11. J.J.Buckley, K.Reilly and L.Jowers: Simulating Continuous Fuzzy Systems II, Information Sciences. To appear.

12. J.J.Buckley, K.Reilly and X.Zheng: Fuzzy Probabilities for Web Planning, Soft Computing, 8(2004)464-476.

13. J.J.Buckley, K.Reilly and X.Zheng: Simulating Fuzzy Systems I, in: Applied Research in Uncertainty Modeling and Analysis, Eds. N.O.Attoh-Okine, B.M.Ayyub, Springer, Heidelberg, Germany, 2005, 31-60.

14. J.J.Buckley, K.Reilly and X.Zheng: Simulating Fuzzy Systems II, in: Applied Research in Uncertainty Modeling and Analysis, Eds. N.O.Attoh-Okine, B.M.Ayyub, Springer, Heidelberg, Germany, 2005, 61-90.

15. A.Klimke, K.Willner and B.Wohlmuth: Uncertainty Modeling Using Fuzzy Arithmetic Based on Sparse Grids: Applications to Dynamic Systems, Int. J. Uncertainty, Fuzziness and Knowledge-Based Systems, 12(2004)745-759.

16. Maple 9, Waterloo Maple Inc., Waterloo, Canada.

17. A.C.Cem Say and A.Kutsi Nircan: Random Generation of Monotonic Functions for Monte Carlo Solutions of Qualitative Differential Equations, Automatica, 41(2005)739-754.

18. Q.Shen and R.R.Leitch: Fuzzy Qualitative Simulation, IEEE Trans. Systems, Man, Cybernetics, 23(1993)1038-1061.

19. www.mathworks.com

Chapter 2

Fuzzy Sets

2.1 Introduction

In this chapter we have collected together the basic ideas from fuzzy sets and fuzzy functions needed for the book. Any reader familiar with fuzzy sets, fuzzy numbers, the extension principle, α-cuts, interval arithmetic, and fuzzy functions may go on and have a look at Section 2.5. In Section 2.5 we briefly discuss fuzzy differential equations. Usually, the only fuzzy differential equations that we have previously investigated were those with fuzzy initial conditions. A good general reference for fuzzy sets and fuzzy logic is [1] and [6].

Our notation specifying a fuzzy set is to place a "bar" over a letter. So $\overline{X}$, $\overline{M}$, $\overline{T}$, ..., $\overline{\mu}$, $\overline{p}$, $\overline{\sigma}^2$, $\overline{a}$, $\overline{b}$, ..., all denote fuzzy sets.

2.2 Fuzzy Sets

If Ω is some set, then a fuzzy subset $\overline{A}$ of Ω is defined by its membership function , written $\overline{A}(x)$, which produces values in $[0, 1]$ for all x in Ω. So, $\overline{A}(x)$ is a function mapping Ω into $[0, 1]$. If $\overline{A}(x_0) = 1$, then we say x_0 belongs to $\overline{A}$, if $\overline{A}(x_1) = 0$ we say x_1 does not belong to $\overline{A}$, and if $\overline{A}(x_2) = 0.6$ we say the membership value of x_2 in $\overline{A}$ is 0.6. When $\overline{A}(x)$ is always equal to one or zero we obtain a crisp (non-fuzzy) subset of Ω. For all fuzzy sets $\overline{B}$, $\overline{C}$, ... we use $\overline{B}(x)$, $\overline{C}(x)$, ... for the value of their membership function at x. The fuzzy sets we will be using will usually be fuzzy numbers .

The term "crisp" will mean not fuzzy. A crisp set is a regular set. A crisp number is just a real number. A crisp function maps real numbers (or real vectors) into real numbers. A crisp solution to a problem is a solution involving crisp sets, crisp numbers, crisp functions, etc.

2.2.1 Fuzzy Numbers

A general definition of fuzzy number may be found in ([1],[6]), however our fuzzy numbers will be triangular (shaped) fuzzy numbers. A triangular fuzzy number $\overline{N}$ is defined by three numbers $a < b < c$ where the base of the triangle is the interval $[a, c]$ and its vertex is at $x = b$. Triangular fuzzy numbers will be written as $\overline{N} = (a/b/c)$. A triangular fuzzy number $\overline{N} = (1.2/2/2.4)$ is shown in Figure 2.1. We see that $\overline{N}(2) = 1$, $\overline{N}(1.6) = 0.5$, etc.

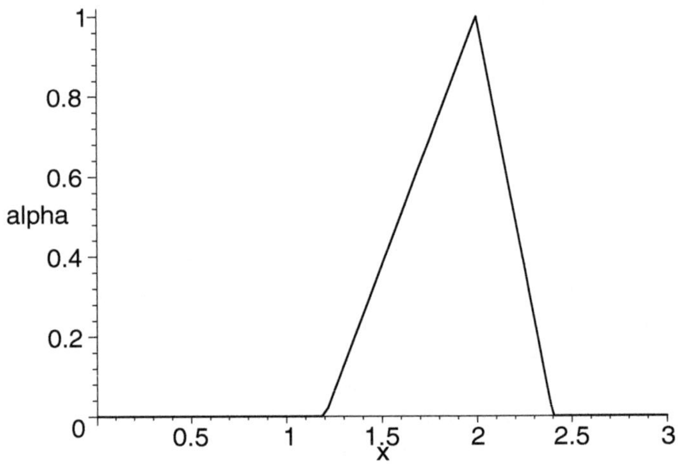

Figure 2.1: Triangular Fuzzy Number $\overline{N}$

A triangular shaped fuzzy number $\overline{P}$ is given in Figure 2.2. $\overline{P}$ is only partially specified by the three numbers 1.2, 2, 2.4 since the graph on $[1.2, 2]$, and $[2, 2.4]$, is not a straight line segment. To be a triangular shaped fuzzy number we require the graph to be continuous and: (1) monotonically increasing on $[1.2, 2]$; and (2) monotonically decreasing on $[2, 2.4]$. For triangular shaped fuzzy number $\overline{P}$ we use the notation $\overline{P} \approx (1.2/2/2.4)$ to show that it is partially defined by the three numbers 1.2, 2, and 2.4. If $\overline{P} \approx (1.2/2/2.4)$ we know its base is on the interval $[1.2, 2.4]$ with vertex (membership value one) at $x = 2$.

2.2.2 Alpha-Cuts

Alpha-cuts are slices through a fuzzy set producing regular (non-fuzzy) sets. If $\overline{A}$ is a fuzzy subset of some set Ω, then an α-cut of $\overline{A}$, written $\overline{A}[\alpha]$, is defined as

$$\overline{A}[\alpha] = \{x \in \Omega | \overline{A}(x) \geq \alpha\}, \tag{2.1}$$

for all α, $0 < \alpha \leq 1$. The $\alpha = 0$ cut, or $\overline{A}[0]$, must be defined separately.

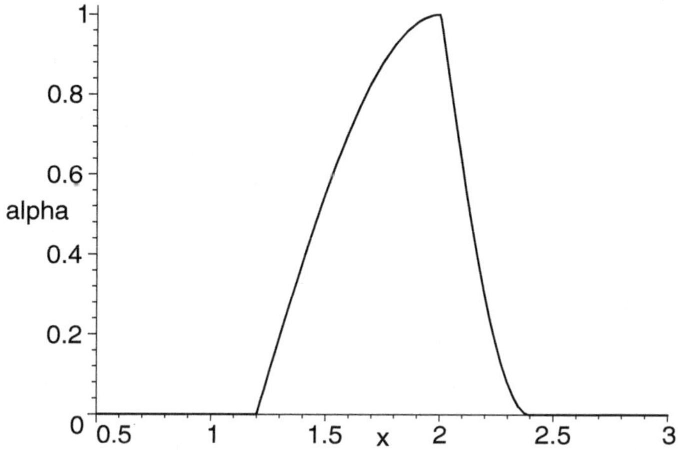

Figure 2.2: Triangular Shaped Fuzzy Number $\overline{P}$

Let $\overline{N}$ be the fuzzy number in Figure 2.1. Then $\overline{N}[0] = [1.2, 2.4]$. Notice that using equation (2.1) to define $\overline{N}[0]$ would give $\overline{N}[0]$ = all the real numbers. Similarly, in Figure 2.2 $\overline{P}[0] = [1.2, 2.4]$. For any fuzzy set $\overline{A}$, $\overline{A}[0]$ is called the support, or base, of $\overline{A}$. Many authors call the support of a fuzzy number the open interval (a, b) like the support of $\overline{N}$ in Figure 2.1 would then be $(1.2, 2.4)$. However in this book we use the closed interval $[a, b]$ for the support (base) of the fuzzy number.

The core of a fuzzy number is the set of values where the membership value equals one. If $\overline{N} = (a/b/c)$, or $\overline{N} \approx (a/b/c)$, then the core of $\overline{N}$ is the single point b.

For any fuzzy number $\overline{Q}$ we know that $\overline{Q}[\alpha]$ is a closed, bounded, interval for $0 \leq \alpha \leq 1$. We will write this as

$$\overline{Q}[\alpha] = [q_1(\alpha), q_2(\alpha)], \tag{2.2}$$

where $q_1(\alpha)$ ($q_2(\alpha)$) will be an increasing (decreasing) function of α with $q_1(1) = q_2(1)$. If $\overline{Q}$ is a triangular shaped then: (1) $q_1(\alpha)$ will be a continuous, monotonically increasing function of α in $[0, 1]$; (2) $q_2(\alpha)$ will be a continuous, monotonically decreasing function of α, $0 \leq \alpha \leq 1$; and (3) $q_1(1) = q_2(1)$.

For the $\overline{N}$ in Figure 2.1 we obtain $\overline{N}[\alpha] = [n_1(\alpha), n_2(\alpha)]$, $n_1(\alpha) = 1.2 + 0.8\alpha$ and $n_2(\alpha) = 2.4 - 0.4\alpha$, $0 \leq \alpha \leq 1$. The equation for $n_i(\alpha)$ is backwards. With the y-axis vertical and the x-axis horizontal the equation $n_1(\alpha) = 1.2 + 0.8\alpha$ means $x = 1.2 + 0.8y$, $0 \leq y \leq 1$. That is, the straight line segment from $(1.2, 0)$ to $(2, 1)$ in Figure 2.1 is given as x a function of y whereas it is usually stated as y a function of x. This is how it will be done for all α-cuts of fuzzy numbers.

The general requirements for a fuzzy set $\overline{N}$ of the real numbers to be a fuzzy number are: (1) it must be normalized, or $\overline{N}(x) = 1$ for some x; and (2) its alpha-cuts must be closed, bounded, intervals for all alpha in $[0,1]$. This will be important in fuzzy estimation because there the fuzzy numbers will have very short vertical line segments at both ends of its base (see Section 3.3 in Chapter 3). Even so, such a fuzzy set still meets the general requirements presented above to be called a fuzzy number.

2.2.3 Inequalities

Let $\overline{N} = (a/b/c)$. We write $\overline{N} \geq \delta$, δ some real number, if $a \geq \delta$, $\overline{N} > \delta$ when $a > \delta$, $\overline{N} \leq \delta$ for $c \leq \delta$ and $\overline{N} < \delta$ if $c < \delta$. We use the same notation for triangular shaped fuzzy numbers whose support is the interval $[a, c]$.

If $\overline{A}$ and $\overline{B}$ are two fuzzy subsets of a set Ω, then $\overline{A} \leq \overline{B}$ means $\overline{A}(x) \leq \overline{B}(x)$ for all x in Ω, or $\overline{A}$ is a fuzzy subset of $\overline{B}$. $\overline{A} < \overline{B}$ holds when $\overline{A}(x) < \overline{B}(x)$, for all x.

2.2.4 Discrete Fuzzy Sets

Let $\overline{A}$ be a fuzzy subset of Ω. If $\overline{A}(x)$ is not zero only at a finite number of x values in Ω, then $\overline{A}$ is called a discrete fuzzy set. Suppose $\overline{A}(x)$ is not zero only at x_1, x_2, x_3 and x_4 in Ω. Then we write the fuzzy set as

$$\overline{A} = \{\frac{\mu_1}{x_1}, \cdots, \frac{\mu_4}{x_4}\}, \tag{2.3}$$

where the μ_i are the membership values. That is, $\overline{A}(x_i) = \mu_i$, $1 \leq i \leq 4$, and $\overline{A}(x) = 0$ otherwise. We can have discrete fuzzy subsets of any space Ω. Notice that α-cuts of discrete fuzzy sets of $\mathbb{R}$, the set of real numbers, do not produce closed, bounded, intervals. We will use a discrete fuzzy set in Chapter 17.

2.3 Fuzzy Arithmetic

If $\overline{A}$ and $\overline{B}$ are two fuzzy numbers we may need to add, subtract, multiply and divide them. There are two basic methods of computing $\overline{A} + \overline{B}$, $\overline{A} - \overline{B}$, etc. which are: (1) extension principle; and (2) α-cuts and interval arithmetic.

2.3.1 Extension Principle

Let $\overline{A}$ and $\overline{B}$ be two fuzzy numbers. If $\overline{A} + \overline{B} = \overline{C}$, then the membership function for $\overline{C}$ is defined as

$$\overline{C}(z) = \sup_{x,y}\{\min(\overline{A}(x), \overline{B}(y)) | x + y = z\} \ . \tag{2.4}$$

If we set $\overline{C} = \overline{A} - \overline{B}$, then

$$\overline{C}(z) = \sup_{x,y}\{\min(\overline{A}(x), \overline{B}(y))|x - y = z\} . \tag{2.5}$$

Similarly, $\overline{C} = \overline{A} \cdot \overline{B}$, then

$$\overline{C}(z) = \sup_{x,y}\{\min(\overline{A}(x), \overline{B}(y))|x \cdot y = z\}, \tag{2.6}$$

and if $\overline{C} = \overline{A}/\overline{B}$,

$$\overline{C}(z) = \sup_{x,y}\{\min(\overline{A}(x), \overline{B}(y))|x/y = z\} . \tag{2.7}$$

In all cases $\overline{C}$ is also a fuzzy number [6]. We assume that zero does not belong to the support of $\overline{B}$ in $\overline{C} = \overline{A}/\overline{B}$. If $\overline{A}$ and $\overline{B}$ are triangular fuzzy numbers then so are $\overline{A} + \overline{B}$ and $\overline{A} - \overline{B}$, but $\overline{A} \cdot \overline{B}$ and $\overline{A}/\overline{B}$ will be triangular shaped fuzzy numbers.

We should mention something about the operator "sup" in equations (2.4)-(2.7). If Ω is a set of real numbers bounded above (there is a M so that $x \leq M$, for all x in Ω), then $\sup(\Omega)$ = the least upper bound for Ω. If Ω has a maximum member, then $\sup(\Omega) = \max(\Omega)$. For example, if $\Omega = [0, 1)$, $\sup(\Omega) = 1$ but if $\Omega = [0, 1]$, then $\sup(\Omega) = \max(\Omega) = 1$. The dual operator to "sup" is "inf". If Ω is bounded below (there is an M so that $M \leq x$ for all $x \in \Omega$), then $\inf(\Omega)$ = the greatest lower bound. For example, for $\Omega = (0, 1]$ $\inf(\Omega) = 0$ but if $\Omega = [0, 1]$, then $\inf(\Omega) = \min(\Omega) = 0$.

Obviously, given $\overline{A}$ and $\overline{B}$, equations (2.4)- (2.7) appear quite complicated to compute $\overline{A} + \overline{B}$, $\overline{A} - \overline{B}$, etc. So, we now present another procedure based on α-cuts and interval arithmetic. First, we present the basics of interval arithmetic.

2.3.2 Interval Arithmetic

We only give a brief introduction to interval arithmetic. For more information the reader is referred to ([7],[8]). Let $[a_1, b_1]$ and $[a_2, b_2]$ be two closed, bounded, intervals of real numbers. If $*$ denotes addition, subtraction, multiplication, or division, then $[a_1, b_1] * [a_2, b_2] = [\alpha, \beta]$ where

$$[\alpha, \beta] = \{a * b | a_1 \leq a \leq b_1, a_2 \leq b \leq b_2\} . \tag{2.8}$$

If $*$ is division, we must assume that zero does not belong to $[a_2, b_2]$. We may simplify equation (2.8) as follows:

$$\begin{aligned} [a_1, b_1] + [a_2, b_2] &= [a_1 + a_2, b_1 + b_2] , & (2.9) \\ [a_1, b_1] - [a_2, b_2] &= [a_1 - b_2, b_1 - a_2] , & (2.10) \\ [a_1, b_1] / [a_2, b_2] &= [a_1, b_1] \cdot \left[\frac{1}{b_2}, \frac{1}{a_2}\right] , & (2.11) \end{aligned}$$

and

$$[a_1, b_1] \cdot [a_2, b_2] = [\alpha, \beta], \tag{2.12}$$

where

$$\alpha = \min\{a_1 a_2, a_1 b_2, b_1 a_2, b_1 b_2\}, \tag{2.13}$$
$$\beta = \max\{a_1 a_2, a_1 b_2, b_1 a_2, b_1 b_2\} . \tag{2.14}$$

Multiplication and division may be further simplified if we know that $a_1 > 0$ and $b_2 < 0$, or $b_1 > 0$ and $b_2 < 0$, etc. For example, if $a_1 \geq 0$ and $a_2 \geq 0$, then

$$[a_1, b_1] \cdot [a_2, b_2] = [a_1 a_2, b_1 b_2], \tag{2.15}$$

and if $b_1 < 0$ but $a_2 \geq 0$, we see that

$$[a_1, b_1] \cdot [a_2, b_2] = [a_1 b_2, a_2 b_1] . \tag{2.16}$$

Also, assuming $b_1 < 0$ and $b_2 < 0$ we get

$$[a_1, b_1] \cdot [a_2, b_2] = [b_1 b_2, a_1 a_2], \tag{2.17}$$

but $a_1 \geq 0$, $b_2 < 0$ produces

$$[a_1, b_1] \cdot [a_2, b_2] = [a_2 b_1, b_2 a_1] . \tag{2.18}$$

2.3.3 Fuzzy Arithmetic

Again we have two fuzzy numbers $\overline{A}$ and $\overline{B}$. We know α-cuts are closed, bounded, intervals so let $\overline{A}[\alpha] = [a_1(\alpha), a_2(\alpha)]$, $\overline{B}[\alpha] = [b_1(\alpha), b_2(\alpha)]$. Then if $\overline{C} = \overline{A} + \overline{B}$ we have

$$\overline{C}[\alpha] = \overline{A}[\alpha] + \overline{B}[\alpha] . \tag{2.19}$$

We add the intervals using equation (2.9). Setting $\overline{C} = \overline{A} - \overline{B}$ we get

$$\overline{C}[\alpha] = \overline{A}[\alpha] - \overline{B}[\alpha], \tag{2.20}$$

for all α in $[0, 1]$. Also

$$\overline{C}[\alpha] = \overline{A}[\alpha] \cdot \overline{B}[\alpha], \tag{2.21}$$

for $\overline{C} = \overline{A} \cdot \overline{B}$ and

$$\overline{C}[\alpha] = \overline{A}[\alpha] / \overline{B}[\alpha], \tag{2.22}$$

when $\overline{C} = \overline{A} / \overline{B}$, provided that zero does not belong to $\overline{B}[\alpha]$ for all α. This method is equivalent to the extension principle method of fuzzy arithmetic [6]. Obviously, this procedure, of α-cuts plus interval arithmetic, is more user (and computer) friendly.

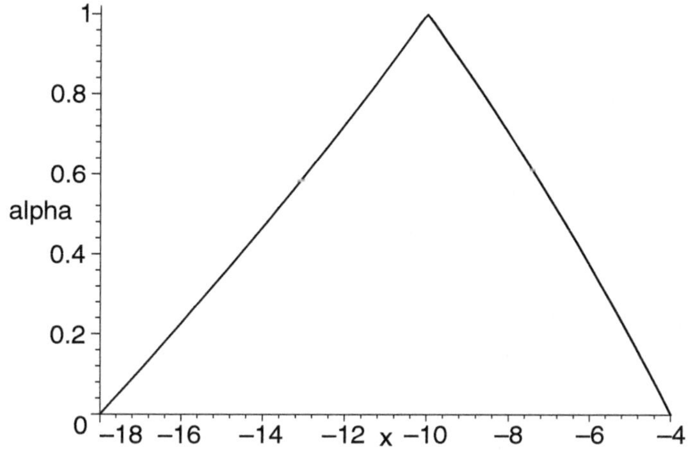

Figure 2.3: The Fuzzy Number $\overline{C} = \overline{A} \cdot \overline{B}$

Example 2.3.3.1

Let $\overline{A} = (-3/-2/-1)$ and $\overline{B} = (4/5/6)$. We determine $\overline{A} \cdot \overline{B}$ using α-cuts and interval arithmetic. We compute $\overline{A}[\alpha] = [-3+\alpha, -1-\alpha]$ and $\overline{B}[\alpha] = [4+\alpha, 6-\alpha]$. So, if $\overline{C} = \overline{A} \cdot \overline{B}$ we obtain $\overline{C}[\alpha] = [(\alpha-3)(6-\alpha), (-1-\alpha)(4+\alpha)]$, $0 \leq \alpha \leq 1$. The graph of $\overline{C}$ is shown in Figure 2.3.

2.4 Fuzzy Functions

In this book a fuzzy function is a mapping from fuzzy numbers into fuzzy numbers. We write $H(\overline{X}) = \overline{Z}$ for a fuzzy function with one independent variable $\overline{X}$. $\overline{X}$ will be a triangular (shaped) fuzzy number and then we usually obtain $\overline{Z}$ as a triangular (shaped) shaped fuzzy number. For two independent variables we have $H(\overline{X}, \overline{Y}) = \overline{Z}$.

Where do these fuzzy functions come from? They are usually extensions of real-valued functions. Let $h : [a,b] \to \mathbb{R}$. This notation means $z = h(x)$ for x in $[a,b]$ and z a real number. One extends $h : [a,b] \to \mathbb{R}$ to $H(\overline{X}) = \overline{Z}$ in two ways: (1) the extension principle; or (2) using α-cuts and interval arithmetic.

2.4.1 Extension Principle

Any $h : [a,b] \to \mathbb{R}$ may be extended to $H(\overline{X}) = \overline{Z}$ as follows

$$\overline{Z}(z) = \sup_x \{ \ \overline{X}(x) \mid h(x) = z, \ a \leq x \leq b \ \} \ . \tag{2.23}$$

Equation (2.23) defines the membership function of $\overline{Z}$ for any triangular (shaped) fuzzy number $\overline{X}$ in $[a, b]$.

If h is continuous, then we have a way to find α-cuts of $\overline{Z}$. Let $\overline{Z}[\alpha] = [z_1(\alpha), z_2(\alpha)]$. Then [3]

$$z_1(\alpha) = \min\{\ h(x) \mid x \in \overline{X}[\alpha]\ \}, \tag{2.24}$$

$$z_2(\alpha) = \max\{\ h(x) \mid x \in \overline{X}[\alpha]\ \}, \tag{2.25}$$

for $0 \le \alpha \le 1$.

If we have two independent variables, then let $z = h(x, y)$ for x in $[a_1, b_1]$, y in $[a_2, b_2]$. We extend h to $H(\overline{X}, \overline{Y}) = \overline{Z}$ as

$$\overline{Z}(z) = \sup_{x,y} \left\{ \min\left(\overline{X}(x), \overline{Y}(y)\right) \mid h(x, y) = z \right\}, \tag{2.26}$$

for $\overline{X}$ ($\overline{Y}$) a triangular (shaped) fuzzy number in $[a_1, b_1]$ ($[a_2, b_2]$). For α-cuts of $\overline{Z}$, assuming h is continuous, we have

$$z_1(\alpha) = \min\{\ h(x, y) \mid x \in \overline{X}[\alpha],\ y \in \overline{Y}[\alpha]\ \}, \tag{2.27}$$

$$z_2(\alpha) = \max\{\ h(x, y) \mid x \in \overline{X}[\alpha],\ y \in \overline{Y}[\alpha]\ \}, \tag{2.28}$$

$0 \le \alpha \le 1$.

2.4.2 Alpha-Cuts and Interval Arithmetic

All the functions we usually use in engineering and science have a computer algorithm which, using a finite number of additions, subtractions, multiplications and divisions, can evaluate the function to required accuracy. Such functions can be extended, using α-cuts and interval arithmetic, to fuzzy functions. Let $h : [a, b] \to \mathbb{R}$ be such a function. Then its extension $H(\overline{X}) = \overline{Z}$, $\overline{X}$ in $[a, b]$ is done, via interval arithmetic, in computing $h(\overline{X}[\alpha]) = \overline{Z}[\alpha]$, α in $[0, 1]$. We input the interval $\overline{X}[\alpha]$, perform the arithmetic operations needed to evaluate h on this interval, and obtain the interval $\overline{Z}[\alpha]$. Then put these α-cuts together to obtain the value $\overline{Z}$. The extension to more independent variables is straightforward.

For example, consider the fuzzy function

$$\overline{Z} = H(\overline{X}) = \frac{\overline{A}\,\overline{X} + \overline{B}}{\overline{C}\,\overline{X} + \overline{D}}, \tag{2.29}$$

for triangular fuzzy numbers $\overline{A}$, $\overline{B}$, $\overline{C}$, $\overline{D}$ and triangular fuzzy number $\overline{X}$ in $[0, 10]$. We assume that $\overline{C} \ge 0$, $\overline{D} > 0$ so that $\overline{C}\,\overline{X} + \overline{D} > 0$. This would be the extension of

$$h(x_1, x_2, x_3, x_4, x) = \frac{x_1 x + x_2}{x_3 x + x_4}\ . \tag{2.30}$$

We would substitute the intervals $\overline{A}[\alpha]$ for x_1, $\overline{B}[\alpha]$ for x_2, $\overline{C}[\alpha]$ for x_3, $\overline{D}[\alpha]$ for x_4 and $\overline{X}[\alpha]$ for x, do interval arithmetic, to obtain interval $\overline{Z}[\alpha]$ for $\overline{Z}$. Alternatively, the fuzzy function

$$\overline{Z} = H(\overline{X}) = \frac{2\overline{X} + 10}{3\overline{X} + 4}, \tag{2.31}$$

would be the extension of

$$h(x) = \frac{2x + 10}{3x + 4} . \tag{2.32}$$

2.4.3 Differences

Let $h : [a, b] \to \mathbb{R}$. Just for this subsection let us write $\overline{Z}^* = H(\overline{X})$ for the extension principle method of extending h to H for $\overline{X}$ in $[a, b]$. We denote $\overline{Z} = H(\overline{X})$ for the α-cut and interval arithmetic extension of h .

We know that $\overline{Z}$ can be different from $\overline{Z}^*$. But for basic fuzzy arithmetic in Section 2.3 the two methods give the same results. In the example below we show that for $h(x) = x(1 - x)$, x in $[0, 1]$, we can get $\overline{Z}^* \neq \overline{Z}$ for some $\overline{X}$ in $[0, 1]$. What is known ([3],[7]) is that for usual functions in science and engineering $\overline{Z}^* \leq \overline{Z}$. Otherwise, there is no known necessary and sufficient conditions on h so that $\overline{Z}^* = \overline{Z}$ for all $\overline{X}$ in $[a, b]$.

There is nothing wrong in using α-cuts and interval arithmetic to evaluate fuzzy functions. Surely, it is user, and computer friendly. However, we should be aware that whenever we use α-cuts plus interval arithmetic to compute $\overline{Z} = H(\overline{X})$ we may be getting something larger than that obtained from the extension principle. The same results hold for functions of two or more independent variables.

Example 2.4.3.1

The example is the simple fuzzy expression

$$\overline{Z} = (1 - \overline{X})\,\overline{X}, \tag{2.33}$$

for $\overline{X}$ a triangular fuzzy number in $[0, 1]$. Let $\overline{X}[\alpha] = [x_1(\alpha), x_2(\alpha)]$. Using interval arithmetic we obtain

$$z_1(\alpha) = (1 - x_2(\alpha))x_1(\alpha), \tag{2.34}$$

$$z_2(\alpha) = (1 - x_1(\alpha))x_2(\alpha), \tag{2.35}$$

for $\overline{Z}[\alpha] = [z_1(\alpha), z_2(\alpha)]$, α in $[0, 1]$.

The extension principle extends the crisp equation $z = (1-x)x$, $0 \leq x \leq 1$, to fuzzy numbers as follows

$$\overline{Z}^*(z) = \sup_x \left\{ \overline{X}(x) | (1 - x)x = z,\ 0 \leq x \leq 1 \right\} . \tag{2.36}$$

Let $\overline{Z}^*[\alpha] = [z_1^*(\alpha), z_2^*(\alpha)]$. Then

$$z_1^*(\alpha) = \min\{(1-x)x | x \in \overline{X}[\alpha]\}, \tag{2.37}$$

$$z_2^*(\alpha) = \max\{(1-x)x | x \in \overline{X}[\alpha]\}, \tag{2.38}$$

for all $0 \leq \alpha \leq 1$. Now let $\overline{X} = (0/0.25/0.5)$, then $x_1(\alpha) = 0.25\alpha$ and $x_2(\alpha) = 0.50 - 0.25\alpha$. Equations (2.34) and (2.35) give $\overline{Z}[0.50] = [5/64, 21/64]$ but equations (2.37) and (2.38) produce $\overline{Z}^*[0.50] = [7/64, 15/64]$. Therefore, $\overline{Z}^* \neq \overline{Z}$. We do know that if each fuzzy number appears only once in the fuzzy expression, the two methods produce the same results ([3],[7]). However, if a fuzzy number is used more than once, as in equation (2.33), the two procedures can give different results.

2.5 Fuzzy Differential Equations

We start off with the second order, linear, constant coefficient ordinary differential equation

$$y'' + ay' + by = g(x) \ , \tag{2.39}$$

for x in interval I. I can be $[0, T]$, for $T > 0$ or I can be $[0, \infty)$. The initial conditions are $y(0) = \gamma_0$, $y'(0) = \gamma_1$. We assume g is continuous on I.

We have usually considered solutions to equation (2.39) only for fuzzy initial conditions $y(0) = \overline{\gamma}_0$, $y'(0) = \overline{\gamma}_1$, for triangular fuzzy numbers $\overline{\gamma}_0$ and $\overline{\gamma}_1$. When there is uncertainty about how the system, governed by equation (2.39), starts off, we model that uncertainty using fuzzy numbers $\overline{\gamma}_0$ and $\overline{\gamma}_1$. This discussion is adapted from [2] and [4]. Those results also contained applications including: (1) an electrical circuit; (2) a vibrating mass; and (3) a dynamic supply and demand model. Later on in [4] we allowed a and b to be fuzzy but with crisp initial conditions. There is no general theory for the case of a and b both fuzzy so those results investigated only two examples: (1) a fuzzy, $\overline{a} > 0$, $b = 0$; and (2) $a = 0$, b fuzzy, $\overline{b} > 0$. In both cases we start off with a homogeneous equation.

We followed the same theme as in other publications involving solving fuzzy equations in that we looked at three different types of solution: $\overline{Y}_c$, $\overline{Y}_e$ and $\overline{Y}_I$. If we fuzzify the crisp equation (2.39) and solve, we are attempting to get what we called the "classical" solution $\overline{Y}_c$. When we first solve equation (2.39) and then fuzzify the crisp solution, using the extension principle, we obtain the extension principle solution $\overline{Y}_e$ and $\overline{Y}_I$ (called the α-cut and interval arithmetic solution) is obtained by extending (fuzzifying) the crisp solution using alpha-cuts and interval arithmetic.

We found that sometimes the classical fuzzy solution does not exist and sometimes $\overline{Y}_e$ and $\overline{Y}_I$ do not solve the original fuzzy differential equation. So there can be problems with these types of solutions. Also, when you fuzzify more of equation (2.39), like a, b, $g(x)$ and the initial conditions, the result gets more complicated and difficult to obtain a precise mathematical expression for the fuzzy solution. However, we do not need to obtain a precise mathematical solution in this book because we will use simulation.

The crisp continuous systems we are interested in will be governed by systems of ordinary differential equations. Many of the parameters in these equations will need to be estimated and to show the uncertainty in the estimator we will use fuzzy number estimators (Chapter 3). We have previously investigated solving systems of linear ordinary differential equations having fuzzy initial conditions [5]. In that paper we only allow for fuzzy initial conditions and we investigate the two solutions $\overline{Y}_c$ and $\overline{Y}_e$. Fuzzifying more parameters in these equations makes the problem too complex for a complete mathematical solution. Three applications were presented: (1) a predator/prey model (also Chapter 7); (2) spread of an infectious disease (Chapter 10); and (3) an arms race model (Chapter 8). In these examples we only fuzzified the initial conditions but in Chapters 7, 8 and 10 other parameters in the models can be estimated and then considered fuzzy. More details about solving fuzzy differential equations is in Section 6.2 of Chapter 6 .

2.6 References

1. J.J.Buckley and E.Eslami: Introduction to Fuzzy Logic and Fuzzy Sets, Physica-Verlag, Heidelberg, Germany, 2002.

2. J.J.Buckley and T.Feuring: Fuzzy Initial Value Problem for Nth Order Linear Differential Equations, Fuzzy Sets and Systems, 121(2001)247-255.

3. J.J.Buckley and Y.Qu: On Using α-Cuts to Evaluate Fuzzy Equations, Fuzzy Sets and Systems, 38(1990)309-312.

4. J.J.Buckley, E.Eslami and T.Feuring: Fuzzy Mathematics in Economics and Engineering, Physica-Verlag, Heidelberg, Germany, 2002.

5. J.J.Buckley, T.Feuring and Y.Hayashi: Linear Systems of First Order Ordinary Differential Equations: Fuzzy Initial Conditions, Soft Computing, 6(2002)415-421.

6. G.J.Klir and B.Yuan: Fuzzy Sets and Fuzzy Logic: Theory and Applications, Prentice Hall, Upper Saddle River, N.J., 1995.

7. R.E.Moore: Methods and Applications of Interval Analysis, SIAM Studies in Applied Mathematics, Philadelphia, 1979.

8. A.Neumaier: Interval Methods for Systems of Equations, Cambridge University Press, Cambridge, U.K., 1990.

Chapter 3

Fuzzy Estimation

3.1 Introduction

In this book we will consider only two methods of fuzzy estimation: (1) expert opinion; and (2) from data using confidence intervals. First we discuss the expert opinion method. Then we look at the confidence interval procedure with particular emphasis on: (1) the fuzzy arrival/service rates; (2) the fuzzy estimator of p = the probability of a "success" in a binomial experiment; and (3) the fuzzy estimator of the mean of a normal distribution when the variance is unknown. More information on fuzzy estimators is in ([1]-[3]).

We will never assume that a parameter in a model is the value of a random variable. Parameter values as values of random variables would put us into the area of stochastic systems of differential equations. Our parameters will always be constants with some of them not having known precise values which then must be estimated by experts or from historical data.

3.2 Expert Opinion

Let us look at an example called the "arms race" model [5], also discussed in detail in Chapter 8. The system of crisp differential equations is

$$\dot{x} = -ax + by + r, \tag{3.1}$$

$$\dot{y} = cx - dy + s, \tag{3.2}$$

where a, b, c, d are all positive constants, r, s are positive, or negative, constants, subject to initial conditions $x(0) = x_0$, $y(0) = y_0$. Here x (y) represents the yearly rates of armament expenditures of nation A (B) in dollars. Consider estimating the constant b. The basic assumption involving b is that the rate of change of x ($\dot{x}$) is directly proportional to the present expenditures of B, which is y. Or, $\dot{x} = by$. How shall we estimate b? Assume we do not have any recent data on these expenditures for these two countries. We turn to expert opinion.

We may obtain a value for the b from some group of experts. This group could consist of only one expert. First assume we have only one expert and he/she is to estimate the value of b. We can solicit this estimate from the expert as is done in estimating job times in project scheduling ([6], Chapter 13). Let b_1 = the "pessimistic" value of b, or the smallest possible value, let b_3 = be the "optimistic" value of b, or the highest possible value, and let b_2 = the most likely value of b. We then ask the expert to give values for b_1, b_2, b_3 and we construct the triangular fuzzy number $\overline{b} = (b_1/b_2/b_3)$ for b. If we have a group of N experts all to estimate the value of b we solicit the b_{1i}, b_{2i} and b_{3i}, $1 \leq i \leq N$, from them. Let b_1 be the average of the b_{1i}, b_2 is the mean of the b_{2i} and b_3 is the average of the b_{3i}. The simplest thing to do is to use $(b_1/b_2/b_3)$ for $\overline{b}$. We now assume, when necessary, this is how we employ expert opinion to obtain fuzzy estimators. This method will be used numerous times in the applications starting in Chapter 7.

3.3 Fuzzy Estimators from Confidence Intervals

Let us next describe the construction of our fuzzy estimators out of a set of confidence intervals computed from data. More details can be found in ([1]-[3]). This type of fuzzy estimator will be used in the applications beginning in Chapter 7. Let X be a random variable with probability density function $f(x;\theta)$ for single parameter θ. Assume that θ is unknown and it must be estimated from a random sample $X_1, ..., X_n$. Let $Y = u(X_1, ..., X_n)$ be a statistic used to estimate θ. Given the values of these random variables $X_i = x_i$, $1 \leq i \leq n$, we obtain a point estimate $\theta^* = y = u(x_1, ..., x_n)$ for θ. We would never expect this point estimate to exactly equal θ so we often also compute a $(1-\beta)100\%$ confidence interval for θ. In this confidence interval one usually sets β equal to 0.10, 0.05 or 0.01.

We propose to find the $(1-\beta)100\%$ confidence interval for all $0.01 \leq \beta < 1$. Starting at 0.01 is arbitrary and you could begin at 0.10 or 0.05 or 0.005, etc. Denote these confidence intervals as

$$[\theta_1(\beta), \theta_2(\beta)], \tag{3.3}$$

for $0.01 \leq \beta < 1$. Add to this the interval $[\theta^*, \theta^*]$ for the 0% confidence interval for θ. Then we have $(1-\beta)100\%$ confidence intervals for θ for $0.01 \leq \beta \leq 1$.

Now place these confidence intervals, one on top of the other, to produce a triangular shaped fuzzy number $\overline{\theta}$ whose α-cuts are the confidence intervals. We have

$$\overline{\theta}[\alpha] = [\theta_1(\alpha), \theta_2(\alpha)], \tag{3.4}$$

for $0.01 \leq \alpha \leq 1$. All that is needed is to finish the "bottom" of $\overline{\theta}$ to make it a complete fuzzy number. We will simply drop the graph of $\overline{\theta}$ straight down to complete its α-cuts so

$$\overline{\theta}[\alpha] = [\theta_1(0.01), \theta_2(0.01)], \tag{3.5}$$

for $0 \leq \alpha < 0.01$. In this way we are using more information in $\overline{\theta}$ than just a point estimate, or just a single interval estimate. Point estimators show no uncertainty in the estimator.

3.3.1 Fuzzy Estimator of μ

Consider X a random variable with probability density function $N(\mu, \sigma^2)$, with unknown mean μ and known variance σ^2. For unknown variance see Section 3.6 and [1]. To estimate μ we obtain a random sample $X_1, ..., X_n$ from $N(\mu, \sigma^2)$. Suppose the mean of this random sample turns out to be $\overline{x}$,which is a crisp number, not a fuzzy number. We know that $\overline{x}$ is $N(\mu, \sigma^2/n)$. So $(\overline{x} - \mu)/(\sigma/\sqrt{n})$ is $N(0, 1)$. Therefore

$$P(-z_{\beta/2} \leq \frac{\overline{x} - \mu}{\sigma/\sqrt{n}} \leq z_{\beta/2}) = 1 - \beta, \tag{3.6}$$

where $z_{\beta/2}$ is the z value so that the probability of a $N(0, 1)$ random variable exceeding it is $\beta/2$. Now solve the inequality for μ producing

$$P(\overline{x} - z_{\beta/2}\sigma/\sqrt{n} \leq \mu \leq \overline{x} + z_{\beta/2}\sigma/\sqrt{n}) = 1 - \beta. \tag{3.7}$$

This leads directly to the $(1 - \beta)100\%$ confidence interval for μ

$$[\theta_1(\beta), \theta_2(\beta)] = [\overline{x} - z_{\beta/2}\sigma/\sqrt{n}, \overline{x} + z_{\beta/2}\sigma/\sqrt{n}], \tag{3.8}$$

where $z_{\beta/2}$ is defined as

$$\int_{-\infty}^{z_{\beta/2}} N(0, 1)dx = 1 - \beta/2, \tag{3.9}$$

and $N(0, 1)$ denotes the normal density with mean zero and unit variance. Put these confidence intervals together as discussed above and we obtain $\overline{\mu}$ our fuzzy estimator of μ.

The following example shows that the fuzzy estimator of the mean of the normal probability density will be a triangular shaped fuzzy number.

Example 3.3.1.1

Consider X a random variable with probability density function $N(\mu, 100)$, which is the normal probability density with unknown mean μ and known

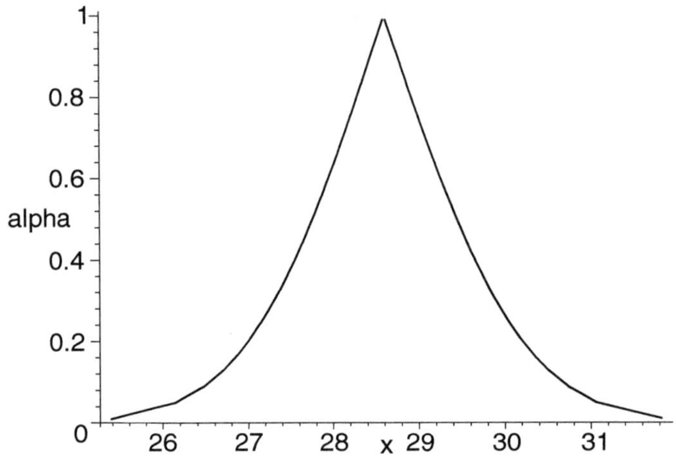

Figure 3.1: Fuzzy Estimator $\overline{\mu}$ in Example 3.3.1.1, $0.01 \leq \beta \leq 1$

variance $\sigma^2 = 100$. To estimate μ we obtain a random sample $X_1, ..., X_n$ from $N(\mu, 100)$. Suppose the mean of this random sample turns out to be 28.6. Then a $(1-\beta)100\%$ confidence interval for μ is

$$[\theta_1(\beta), \theta_2(\beta)] = [28.6 - z_{\beta/2}10/\sqrt{n}, 28.6 + z_{\beta/2}10/\sqrt{n}]. \tag{3.10}$$

To obtain a graph of fuzzy μ, or $\overline{\mu}$, let $n = 64$ and assume that $0.01 \leq \beta \leq 1$. We evaluated equation (3.10) using Maple [4] and then the final graph of $\overline{\mu}$ is shown in Figure 3.1, without dropping the graph straight down to the x-axis at the end points.

For simplicity we will use triangular fuzzy numbers, instead of triangular shaped fuzzy numbers, for fuzzy estimators in the rest of the book.

Now we concentrate on some specific fuzzy estimators to be used in the book.

3.4 Fuzzy Arrival/Service Rates

In this section we concentrate on deriving fuzzy numbers for the arrival rate, and the service rate, in a queuing system. We consider the fuzzy arrival rate first.

3.4.1 Fuzzy Arrival Rate

We assume that we have Poisson arrivals ([6], Chapter 15) which means that there is a positive constant λ so that the probability of k arrivals per unit time is

$$\lambda^k \exp(-\lambda)/k!, \tag{3.11}$$

the Poisson probability function. We need to estimate λ, the arrival rate, so we take a random sample $X_1, ..., X_m$ of size m. In the random sample X_i is the number of arrivals per unit time, in the ith observation. Let S be the sum of the X_i and let $\overline{X}$ be S/m. Here, $\overline{X}$ is not a fuzzy set but the mean.

Now S is Poisson with parameter $m\lambda$ ([7], p. 298). Assuming that $m\lambda$ is sufficiently large (say, at least 30), we may use the normal approximation ([7], p. 317), so the statistic

$$W = \frac{S - m\lambda}{\sqrt{m\lambda}}, \tag{3.12}$$

is approximately a standard normal. Then

$$P[-z_{\beta/2} < W < z_{\beta/2}] \approx 1 - \beta, \tag{3.13}$$

where the $z_{\beta/2}$ was defined in equation (3.9). Now divide numerator and denominator of W by m and we get

$$P[-z_{\beta/2} < Z < z_{\beta/2}] \approx 1 - \beta, \tag{3.14}$$

where

$$Z = \frac{\overline{X} - \lambda}{\sqrt{\lambda/m}}. \tag{3.15}$$

From these last two equations we may derive an approximate $(1-\beta)100\%$ confidence interval for λ. Let us call this confidence interval $[l(\beta), r(\beta)]$.

We now show how to compute $l(\beta)$ and $r(\beta)$. Let

$$f(\lambda) = \sqrt{m}(\overline{X} - \lambda)/\sqrt{\lambda}. \tag{3.16}$$

Now $f(\lambda)$ has the following properties: (1) it is strictly decreasing for $\lambda > 0$; (2) it is zero for $\lambda > 0$ only at $\overline{X} = \lambda$; (3) the limit of f, as λ goes to ∞ is $-\infty$; and (4) the limit of f as λ approaches zero from the right is ∞. Hence, (1) the equation $z_{\beta/2} = f(\lambda)$ has a unique solution $\lambda = l(\beta)$; and (2) the equation $-z_{\beta/2} = f(\lambda)$ also has a unique solution $\lambda = r(\beta)$.

We may find these unique solutions. Let

$$V = \sqrt{z_{\beta/2}^2/m + 4\overline{X}}, \tag{3.17}$$

$$z_1 = [-\frac{z_{\beta/2}}{\sqrt{m}} + V]/2, \tag{3.18}$$

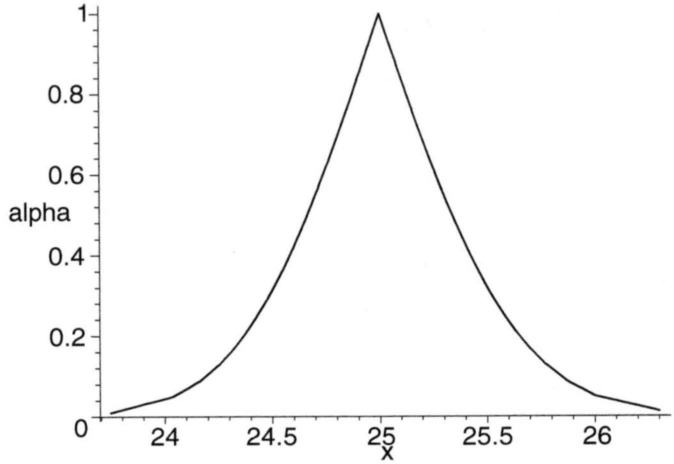

Figure 3.2: Fuzzy Arrival Rate $\overline{\lambda}$ in Example 3.4.1.1

and

$$z_2 = [\frac{z_{\beta/2}}{\sqrt{m}} + V]/2. \tag{3.19}$$

Then $l(\beta) = z_1^2$ and $r(\beta) = z_2^2$.

We now substitute α for β to get the α-cuts of fuzzy number $\overline{\lambda}$. Add the point estimate, when $\alpha = 1$, $\overline{X}$, for the 0% confidence interval. Now as α goes from 0.01 (99% confidence interval) to one (0% confidence interval) we get the fuzzy number for λ. As before, we drop the graph straight down at the ends to obtain a complete fuzzy number.

Example 3.4.1.1

Suppose $m = 100$ and we obtained $\overline{X} = 25$. We evaluated equations (3.17) through (3.19) using Maple [4] and then the graph of $\overline{\lambda}$ is shown in Figure 3.2, without dropping the graph straight down to the $x-$axis at the end points. However, in the rest of the book we will use a triangular fuzzy number for $\overline{\lambda}$.

3.4.2 Fuzzy Service Rate

Let μ be the average (expected) service rate, in the number of service completions per unit time, for a busy server. Then $1/\mu$ is the average (expected) service time. The probability density of the time interval between successive service completions is ([6], Chapter 13)

$$(1/\mu)\exp(-t/\mu), \tag{3.20}$$

for $t > 0$, the exponential probability density function. Let $X_1, ..., X_n$ be a random sample from this exponential density function. Then the maximum likelihood estimator for μ is $\overline{X}$ ([7],p.344), the mean of the random sample (not a fuzzy set). We know that the probability density for $\overline{X}$ is the gamma ([7],p.297) with mean μ and variance μ^2/n ([7],p.351). If n is sufficiently large we may use the normal approximation to determine approximate confidence intervals for μ. Let

$$Z = (\sqrt{n}[\overline{X} - \mu])/\mu, \tag{3.21}$$

which is approximately normally distributed with zero mean and unit variance, provided n is sufficiently large. See Figure 6.4-2 in [7] for $n = 100$ which shows the approximation is quite good if $n = 100$. The graph in Figure 6.4-2 in [7] is for the chi-square distribution which is a special case of the gamma distribution. So we now assume that $n \geq 100$ and use the normal approximation to the gamma.

An approximate $(1 - \beta)100\%$ confidence interval for μ is obtained from

$$P[-z_{\beta/2} < Z < z_{\beta/2}] \approx 1 - \beta, \tag{3.22}$$

where $z_{\beta/2}$ was defined in equation (3.9). After solving for μ we get

$$P[L(\beta) < \mu < R(\beta)] \approx 1 - \beta, \tag{3.23}$$

where

$$L(\beta) = [\sqrt{n}\,\overline{X}]/[z_{\beta/2} + \sqrt{n}], \tag{3.24}$$

and

$$R(\beta) = [\sqrt{n}\,\overline{X}]/[\sqrt{n} - z_{\beta/2}]. \tag{3.25}$$

An approximate $(1 - \beta)100\%$ confidence interval for μ is

$$[\frac{\sqrt{n}\,\overline{X}}{z_{\beta/2} + \sqrt{n}}, \frac{\sqrt{n}\,\overline{X}}{\sqrt{n} - z_{\beta/2}}]. \tag{3.26}$$

Example 3.4.2.1

If $n = 400$ and $\overline{X} = 1.5$, then we get

$$[\frac{30}{z_{\beta/2} + 20}, \frac{30}{20 - z_{\beta/2}}], \tag{3.27}$$

for a $(1 - \beta)100\%$ confidence interval for the service rate μ. Now we can put these confidence intervals together, one on top of another, to obtain a fuzzy number $\overline{\mu}$ for the service rate. We evaluated equation (3.27) using Maple [4] for $0.01 \leq \beta \leq 1$ and the graph of the fuzzy service rate, without dropping the graph straight down to the x-axis at the end points, is in Figure 3.3. For simplicity we use triangular fuzzy numbers for $\overline{\mu}$ in the rest of the book.

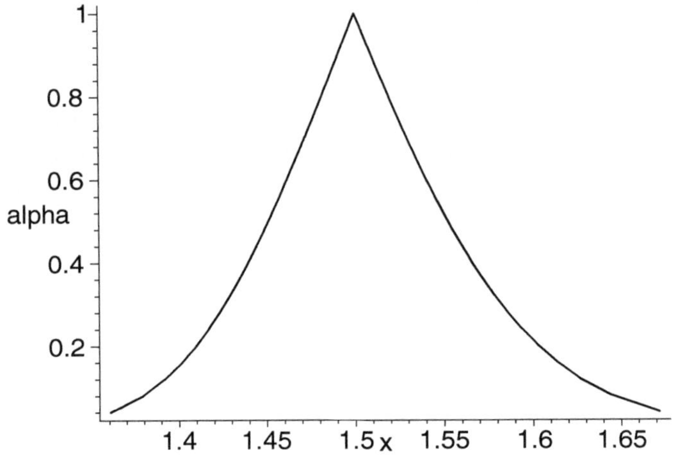

Figure 3.3: Fuzzy Service Rate $\overline{\mu}$ in Example 3.4.2.1

3.5 Fuzzy Estimator of p in the Binomial

We have an experiment in mind in which we are interested in only two possible outcomes labeled "success" and "failure". Let p be the probability of a success so that $q = 1 - p$ will be the probability of a failure. We want to estimate the value of p. We therefore gather a random sample which here is running the experiment n independent times and counting the number of times we had a success. Let x be the number of times we observed a success in n independent repetitions of this experiment. Then our point estimate of p is $\widehat{p} = x/n$.

We know that (Section 7.5 in [7]) that $(\widehat{p} - p)/\sqrt{p(1-p)/n}$ is approximately $N(0,1)$ if n is sufficiently large. Throughout this book we will always assume that the sample size is large enough for the normal approximation to the binomial. Then

$$P(z_{\beta/2} \leq \frac{\widehat{p} - p}{\sqrt{p(1-p)/n}} \leq z_{\beta/2}) \approx 1 - \beta, \tag{3.28}$$

where $z_{\beta/2}$ was defined in equation (3.9). Solving the inequality for the p in the numerator we have

$$P(\widehat{p} - z_{\beta/2}\sqrt{p(1-p)/n} \leq p \leq \widehat{p} + z_{\beta/2}\sqrt{p(1-p)/n}) \approx 1 - \beta. \tag{3.29}$$

This leads to the $(1-\beta)100\%$ approximate confidence interval for p

$$[\widehat{p} - z_{\beta/2}\sqrt{p(1-p)/n}, \widehat{p} + z_{\beta/2}\sqrt{p(1-p)/n}]. \tag{3.30}$$

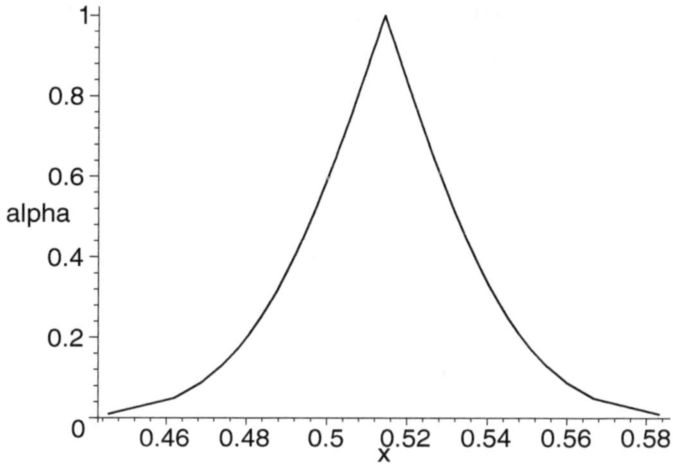

Figure 3.4: Fuzzy Estimator $\overline{p}$ in Example 3.5.1

However, we have no value for p to use in this confidence interval. So, still assuming that n is sufficiently large, we substitute $\widehat{p}$ for p in equation (3.30), using $\widehat{q} = 1 - \widehat{p}$, and we get the final $(1-\beta)100\%$ approximate confidence interval

$$[\widehat{p} - z_{\beta/2}\sqrt{\widehat{p}\widehat{q}/n}, \widehat{p} + z_{\beta/2}\sqrt{\widehat{p}\widehat{q}/n}]. \tag{3.31}$$

Put these confidence intervals together, as discussed above, and we get $\overline{p}$ our triangular shaped fuzzy number estimator of p.

Example 3.5.1

Assume that $n = 350$, $x = 180$ so that $\widehat{p} = 0.5143$. The confidence intervals become

$$[0.5143 - 0.0267z_{\beta/2}, 0.5143 + 0.0267z_{\beta/2}], \tag{3.32}$$

for $0.01 \leq \beta \leq 1$.

We evaluated equation (3.32) using Maple [4] and then the graph of $\overline{p}$ is shown in Figure 3.4, without dropping the graph straight down to the x-axis at the end points. The base ($\overline{\mu}[0]$) in Figure 3.4 is an approximate 99% confidence interval for p.

3.6 Fuzzy Estimator of the Mean of the Normal Distribution

Consider X a random variable with probability density function $N(\mu, \sigma^2)$, which is the normal probability density with unknown mean μ and unknown variance σ^2. To estimate μ we obtain a random sample $X_1, ..., X_n$ from $N(\mu, \sigma^2)$. Suppose the mean of this random sample turns out to be $\overline{x}$, which is a crisp number, not a fuzzy number. Also, let s^2 be the sample variance. Our point estimator of μ is $\overline{x}$. If the values of the random sample are $x_1, ..., x_n$ then the expression we will use for s^2 in this book is

$$s^2 = \sum_{i=1}^{n} (x_i - \overline{x})^2/(n-1). \tag{3.33}$$

We will use this form of s^2, with denominator $(n-1)$, so that it is an unbiased estimator of σ^2.

It is known that $(\overline{x} - \mu)/(s/\sqrt{n})$ has a (Student's) t distribution with $n-1$ degrees of freedom (Section 7.2 of [7]). It follows that

$$P(-t_{\beta/2} \leq \frac{\overline{x} - \mu}{s/\sqrt{n}} \leq t_{\beta/2}) = 1 - \beta, \tag{3.34}$$

where $t_{\beta/2}$ is defined from the (Student's) t distribution, with $n-1$ degrees of freedom, so that the probability of exceeding it is $\beta/2$. Now solve the inequality for μ giving

$$P(\overline{x} - t_{\beta/2}s/\sqrt{n} \leq \mu \leq \overline{x} + t_{\beta/2}s/\sqrt{n}) = 1 - \beta. \tag{3.35}$$

For this we immediately obtain the $(1-\beta)100\%$ confidence interval for μ

$$[\overline{x} - t_{\beta/2}s/\sqrt{n}, \overline{x} + t_{\beta/2}s/\sqrt{n}]. \tag{3.36}$$

Put these confidence intervals together, as discussed before, and we obtain $\overline{\mu}$ our fuzzy number estimator of μ.

Example 3.6.1

Consider X a random variable with probability density function $N(\mu, \sigma^2)$, which is the normal probability density with unknown mean μ and unknown variance σ^2. To estimate μ we obtain a random sample $X_1, ..., X_n$ from $N(\mu, \sigma^2)$. Suppose the mean of this random sample of size 25 turns out to be 28.6 and $s^2 = 3.42$. Then a $(1-\beta)100\%$ confidence interval for μ is

$$[28.6 - t_{\beta/2}\sqrt{3.42/25}, 28.6 + t_{\beta/2}\sqrt{3.42/25}]. \tag{3.37}$$

We evaluated equation (3.37) using Maple [4] and then the graph of $\overline{\mu}$ is shown in Figure 3.5, without dropping the graph straight down to the x-axis at the end points.

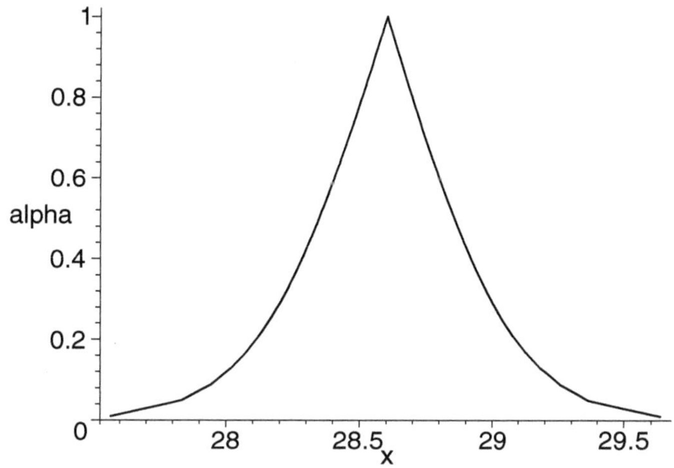

Figure 3.5: Fuzzy Estimator $\overline{\mu}$ in Example 3.6.1

3.7 Summary

We saw in this chapter that our fuzzy estimators can be triangular fuzzy numbers; or triangular shaped fuzzy numbers where we complete the base by drawing short vertical line segments from the horizontal axis up to the graph and the base represents a 99% confidence interval. In the rest of this book, for simplicity, all our fuzzy estimators will be triangular fuzzy numbers.

3.8 References

1. J.J.Buckley: Fuzzy Statistics, Springer, Heidelberg, Germany, 2004.

2. J.J.Buckley: Fuzzy Probabilities and Fuzzy Sets for Web Planning, Springer, Heidelberg, Germany, 2004.

3. J.J.Buckley: Simulating Fuzzy Systems, Springer, Heidelberg, Germany, 2005.

4. Maple 9, Waterloo Maple Inc., Waterloo, Canada.

5. M.Olinick: An Introduction to Mathematical Models in the Social and Life Sciences, Addison-Wesley, Reading, MA, 1978.

6. H.A.Taha: Operations Research, Fifth Edition, Macmillan, N.Y., 1992.

7. R.V.Hogg and E.A.Tanis: Probability and Statistical Inference, Sixth Edition, Prentice Hall, Upper Saddle River, N.J., 2001.

Chapter 4

Fuzzy Systems

4.1 Introduction

We have argued before that many crisp systems become fuzzy systems because the values of some of the parameters in the crisp system are not known precisely and need to be estimated. Too often point estimators for these parameters are obtained and used in the models. But the point estimators do not contain any uncertainty. We proposed incorporating uncertainty into these estimators by employing fuzzy estimators discussed in Chapter 3. Fuzzy estimators are fuzzy numbers so some of the parameters in the model become fuzzy numbers making it into a fuzzy system. We will now discuss this transformation from crisp system into a fuzzy system in more detail for continuous systems.

This book is concerned with using crisp continuous simulation to estimate the evolution of continuous fuzzy systems. The continuous fuzzy systems we will look at are all governed by fuzzy differential equations. So, let us first look at our previous work in this area of using crisp simulation to study fuzzy systems([1],[2],[6]-[9]).

We started with studying what are called discrete event (fuzzy) systems. These systems can usually be described as queuing networks. Items (transactions) arrive at various points in the system and go into a queue waiting for service. The service stations, preceded by a queue, are connected forming a network of queues and service, until the transaction finally exits the system. Examples considered included machine shops, emergency rooms, project networks, bus routes, etc. Analysis of all of these systems depends on parameters like arrival rates and service rates. These parameters are usually estimated from historical data. These estimators are generally point estimators. The point estimators are put into the model to compute system descriptors like mean time an item spends in the system, or the expected number of transactions leaving the system per unit time. We argued that these point estimators contain uncertainty not shown in the calculations. Our estimators of these parameters become fuzzy numbers constructed by placing a set of confidence intervals one on top of another (Chapter 3). Using fuzzy number parameters in the model makes it into a fuzzy system. The system descriptors we want (time in system, number leaving per unit time) will be fuzzy numbers. In general computing these fuzzy numbers can be difficult. We showed how crisp discrete event simulation can be used to estimate the fuzzy numbers used to describe system behavior.

Continuous systems are usually described by a system of ordinary differential equations (ODEs). Many parameters in the system of ODEs are not known precisely and must be estimated. To show the uncertainty in these parameter values we will use fuzzy number estimators. Fuzzy number parameter values produce a system of fuzzy ODEs to solve and we have a continuous fuzzy system. Solution trajectories become fuzzy trajectories. We plan to use crisp continuous simulation to estimate these fuzzy trajectories.

We will start with a continuous crisp system whose description in time depends on crisp ordinary differential equations. Let us consider an example of a predator/prey model, also discussed in detail in Chapter 7. This is adopted from an example in ([10],[11]). The system of differential equations is

$$\dot{x} = -ax + bxy, \tag{4.1}$$

$$\dot{y} = dy - cxy, \tag{4.2}$$

for constants a, b, c, d all positive and initial conditions $x(0) = x_0$, $y(0) = y_0$. We write the time derivative of x (y) as $\dot{x}$ ($\dot{y}$). Non-trivial solutions to this system can not be obtained in terms of elementary functions. So we would need to employ some software to obtain the graphs of $x(t)$ and $y(t)$. How do we get values for a, b, c, d? We considered two cases in in Chapter 3: (1) their values are estimated by expert opinion; and (2) their values are estimated from data by placing confidence intervals one on top of another. In either case the estimators become fuzzy, or fuzzy numbers.

In the predator/prey model above we could have some, or all the parameters, fuzzy. Therefore, $\overline{a}$, $\overline{b}$, $\overline{c}$, $\overline{d}$, $\overline{x}_0$ and $\overline{y}_0$ may all be fuzzy. Then we have a system of fuzzy nonlinear differential equations to solve. The trajectories for $\overline{x}(t)$ and $\overline{y}(t)$, $t \geq 0$, will be fuzzy which means for each value of t $\overline{x}(t)$ and $\overline{y}(t)$ will be fuzzy numbers. We have considered solving fuzzy differential equations before ([3]-[5]). However, in those papers/book we almost always allowed for only fuzzy initial conditions because the fuzzy solution became too difficult to obtain when more parameters became fuzzy. Now we may fuzzify more parameters because we are not finding a formula for the mathematical solution but instead we will use simulation.

4.2 Fuzzy System

Continuing with the predator/prey model now assume that through expert opinion (Chapter 3) we have obtained fuzzy estimators $\overline{a}$, $\overline{b}$, $\overline{c}$, $\overline{d}$ for a, b, c, d, respectively. Also assume that the initial values for x and y are known, crisp and not fuzzy. We have a fuzzy system described by fuzzy differential equations

$$\dot{x} = -\overline{a}x + \overline{b}xy, \tag{4.3}$$

$$\dot{y} = \overline{d}y - \overline{c}xy, \tag{4.4}$$

with $x(0) = x_0$, $y(0) = y_0$. Let the solutions be $\overline{x}(t)$ and $\overline{y}(t)$. What we shall want is the fuzzy trajectories $\overline{x}(t)[0]$ and $\overline{y}(t)[0]$, or the alpha zero cut of the fuzzy solutions. The $\alpha = 0$ cut gives us the maximum uncertainty in the solution due to the uncertainty in some of the parameter values.

Let $\overline{x}(t)[\alpha] = [x_1(t,\alpha), x_2(t,\alpha)]$ and $\overline{y}(t)[\alpha] = [y_1(t,\alpha), y_2(t,\alpha)]$, for $t \geq 0$ and $0 \leq \alpha \leq 1$. The bounds on the alpha zero cut of $\overline{x}(t)$ ($\overline{y}(t)$) are given by the functions $x_i(t,0)$ ($y_i(t,0)$), $i = 1,2$.

So to describe the fuzzy system we need to find the functions $x_i(t,0)$ and $y_i(t,0)$ and then $x_2(t,0) - x_1(t,0)$ and $y_2(t,0) - y_1(t,0)$ are the uncertainty bands (intervals) on the (crisp) solutions due to the uncertainty in some of the parameters in the model. If we let a, b, c, d wander throughout the $\alpha = 0$ cut of $\overline{a}$, $\overline{b}$, $\overline{c}$, $\overline{d}$ respectively, then crisp solutions to the predator/prey model will wander within the uncertainty band.

4.3 Computing the Uncertainty Band

Choosing values for the parameters in their alpha zero cut to get the $x_i(t,0)$ and $y(t,0)$, $i = 1,2$, is what we call the "Simulation Optimization" problem to be discussed in Chapter 6, and in all the applications starting in Chapter 7.

We will use continuous simulation software to graph solutions to the crisp system of differential equations. Selection of the software to be used in this book is discussed in Chapter 5.

Quite often the selection of parameter values in their alpha zero cut to get say $x_1(t,0)$ and $x_2(t,0)$ is what we call "intuitive", or obvious. Other times this can not be done. When it is not clear how to pick the values of the fuzzy parameters in their $\alpha = 0$ cut to obtain $x_i(t,0)$ we try combinations to approximate $x_i(t,0)$ and $y_i(t,0)$. More details are in Chapter 7.

Suppose that we could solve the crisp predator/prey system of differential equations giving the precise mathematical solutions

$$x(t) = F(t; a, b, c, d, x_0, y_0), \tag{4.5}$$

and

$$y(t) = G(t; a, b, c, d, x_0, y_0). \tag{4.6}$$

The solutions will be functions of time t and all the parameters. Now the initial values of x and y are crisp and fixed. Next assume that for each value of $t \geq 0$ we can show that F is: (1) increasing in a and d; and (2) decreasing in b and c. Let $\overline{a}[0] = [a_1, a_2]$, $\overline{b}[0] = [b_1, b_2]$, $\overline{c}[0] = [c_1, c_2]$ and $\overline{d}[0] = [d_1, d_2]$. Then we would know how to get $x_1(t, 0)$: (1) choose $a = a_1$ and $d = d_1$; and (2) choose $b = b_2$ and $c = c_2$. For $x_2(t, 0)$: (1) pick $a = a_2$,$d = d_2$; and (2) pick $b = b_1$, $c = c_1$. Similarly, we may choose the parameters for the $y_i(t, 0)$. However, in the continuous system models that we are interested in we do not have closed form solutions (the F and G) available, or the solutions are to complicated to accomplish the above analysis. That is why we will employ continuous simulation software to get the graphs of the crisp solutions.

4.4 Uncertainty Band as a Confidence Band

Suppose all the uncertain parameters in our model are estimated from data as in Sections 3.3 - 3.6 of Chapter 3. The bases of all the fuzzy estimators are 99% confidence intervals. As we allow the parameters to range through their 99% confidence intervals we generate the uncertainty band $\overline{x}(t)[0]$ for some solution $x(t)$. But now this uncertainty band is like a 99% confidence band for the solutions. $\overline{x}(t)[0]$ is like a band of 99% confidence intervals for $x(t)$. We can not make this claim if some of the uncertain parameters are estimated by experts (Section 3.2) because the base of such a fuzzy estimators is not a 99% confidence interval.

4.5 References

1. J.J.Buckley: Simulating Fuzzy Systems, Springer, Heidelberg, Germany, 2005.

2. J.J.Buckley: Fuzzy Systems, Soft Computing. To appear.

3. J.J.Buckley and T.Feuring: Fuzzy Initial Value Problem for Nth Order Linear Differential Equations, Fuzzy Sets and Systems, 121(2001)247-255.

4. J.J.Buckley, E.Eslami and T.Feuring: Fuzzy Mathematics in Economics and Engineering, Physica-Verlag, Heidelberg, Germany, 2002.

5. J.J.Buckley, T.Feuring and Y.Hayashi: Linear Systems of First Order Ordinary Differential Equations: Fuzzy Initial Conditions, Soft Computing, 6(2002)415-421.

6. J.J.Buckley, K.Reilly and L.Jowers: Simulating Continuous Fuzzy Systems I, Iranian J. Fuzzy Systems. To appear.

7. J.J.Buckley, K.Reilly and L.Jowers: Simulating Continuous Fuzzy Systems II, Information Sciences. To appear.

8. J.J.Buckley, K.Reilly and X.Zheng: Simulating Fuzzy Systems I in: Applied Research in Uncertainty Modeling and Analysis, Eds. N.O.Attoh-Okine,B.M.Ayyub, Springer, Heidelberg, Germany, 2005, 31-60.

9. J.J.Buckley, K.Reilly and X.Zheng: Simulating Fuzzy Systems II in: Applied Research in Uncertainty Modeling and Analysis, Eds. N.O.Attoh-Okine,B.M.Ayyub, Springer, Heidelberg, Germany, 2005, 61-90.

10. M.R.Spiegel: Applied Differential Equations, Third Edition, Prentice-Hall, Englewood Cliffs, NJ, 1981.

11. D.G.Zill: A First Course in Differential Equations, Brooks/Cole, Pacific Grove, CA, 1997.

Chapter 5

Continuous Simulation Software

5.1 Software Selection

Now we come to the point were we need to select simulation software to do all the crisp continuous simulations starting in Chapter 7. The authors are not experts in continuous simulation and they did not know about all the products that are available. They decided to do a "search" for continuous simulation products on the web.

In choosing a simulation package we have these main constraints: (1) it must be inexpensive, hopefully at most 100 US dollars; (2) it must be easy to use; (3) it has to run on a desktop computer; and (4) the figures created by the software must be exportable to LaTeX (the word processor we would be using). To start the search we put "continuous simulation" into our web search engine and start looking through the pages. There are hundreds of simulation software packages for educational use [4]. We did come up with the IEEE's "Buyer's Guide Simulation Software" [2] (though not available to the public) which contains most of the software companies marketing continuous simulation.

We also want the simulation software self contained and ready to use so we do not need to write code that requires additional development tools (C++, Java,...) to run a simulation. Obviously, the simulation software package must have a good user's manual. We narrowed in on "click-drag-drop" simulation software. The "click-drag-drop" method makes it very easy to construct a simulation network diagram. Each block for such a system has a corresponding icon in its library. One clicks on the icon in the library (generally this is by depressing the left mouse button). Holding the icon, one positions the icon in the desired location in the network. Finally, the icon is dropped in place (at the release of the mouse button). One connects the

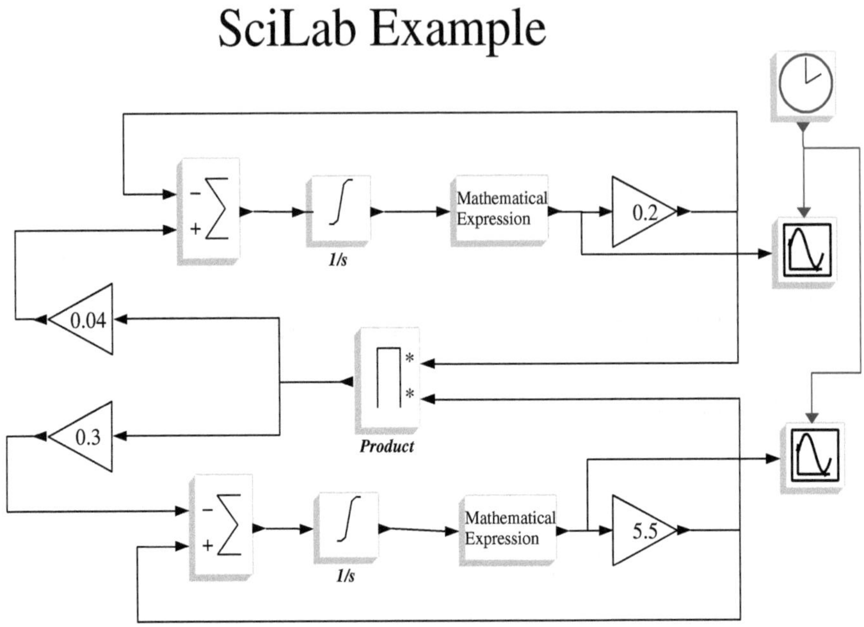

Figure 5.1: SciLab SciCos Diagram

icons using the mouse. Examples of these networks are in all the application chapters. Separate windows open for one to modify information about particular icons. For example, an "Integration" icon (operation) has an embedded parameter for the initial value of the integration output. We found this the easiest method to build the systems we want to simulate. Our search was narrowed down to The MathWorks' Matlab & Simulink [3]. However, this product is expensive.

Consequently, we searched for Simulink clones. We did find a product similar to Simulink, SciCos in SciLab, that can be downloaded for free [1]. We used this continuous simulation package to build a few small continuous systems and then fuzzify them; that is, allow many of the parameters to be fuzzy numbers. Although we found the SciLab clone to be functional, and on recent review we are pleased to note some improvements, its user interface was significantly more difficult than expected. Figure 5.1 is an example of a SciCos diagram which one may compare with the Simulink diagram of Chapter 7. In doing this we found out that this product had limited documentation and we would be unable to use it, in general, for larger systems.

Most simulation software comes in two types: (1) a "scaled down" version sometimes called the "student version", "academic version" or "trial version"; and (2) the full version usually called the "standard" or "professional" version. The professional software of Matlab is quite expensive, usually in the thousands of dollars, so we were interested in using the student software. One may be able to download the student (and professional) software and use it for free for a short time period. We evaluated a trial version; however, it would be a disaster to have the free time period expire in the middle of writing this book. The student version of Matlab & Simulink is a restricted license; use for commercial purposes is prohibited. Fortunately, the authors discovered, our university's Computer and Information Sciences Department already has a licensing agreement for Matlab (needed for Simulink); hence, all our continuous simulations could be performed (with no extra cost) with the professional version of Simulink.

Our choice of Simulink only partially met our selection criteria, but we believe it to be the correct choice. No attempt has been made to guarantee that our simulations will execute on any of the trial Simulink, student Simulink, or clones. We suspect that they are implementable but not necessarily portable to the SciLab clone. We found that: (1) Matlab profession is expensive in our opinion; the student version is a little over 100USD; Scilab is free; (2) Simulink is easy to use, but Matlab programming is required to effect the simulations we present; (3) Matlab & Simulink run on a desktop computer; and (4) the figures created by the software may be exported as Encapulated PostScript (EPS) or color EPS, both of which generate scalable figures for most word processors (including LaTeX). The reason for the late use of Simulink at the authors' university was that the first author, who initially started the book, is in the Mathematics Department; once the CIS department made the most recent release of Matlab available, our selection was complete. We will now use Simulink for all our crisp continuous simulations.

5.2 References

1. http://scilabsoft.inria.fr/

2. http://solutions.iienet.org/

3. http://www.mathworks.com/

4. http://www.sosresearch.org/simulationeducation/simsoftware.html

8.2 References

Chapter 6

Simulation Optimization

6.1 Introduction

Let $x(t)$ be the solution for a variable, in the crisp continuous system, that we wish to study. Fuzzy estimation of parameters in the system leads to a fuzzy system and a fuzzy trajectory $\overline{x}(t)$. For each $t \geq 0$, $\overline{x}(t)$ is a fuzzy number. Let $\overline{x}(t)[\alpha] = [x_1(t,\alpha), x_2(t,\alpha)]$, for $0 \leq \alpha \leq 1$. We want to look at the alpha equal zero graph

$$(x_1(t,0), x_2(t,0)), \ 0 \leq t \leq t_m, \tag{6.1}$$

where t_m is some maximum time. This graph shows the maximum uncertainty in the trajectory of x due to the uncertainty in the values of some of the parameters in the model. Let us now look at some examples of finding the $x_i(t,0)$, $i = 1,2$. In the first example it is easy to find the $x_i(t,0)$ but in the second example finding these functions is more difficult. All of this discussion is what we call the "naive" approach that we used successfully before [1] in discrete event fuzzy systems. However, this naive method does not necessarily work for continuous fuzzy systems. So, in the second section we look in more detail at the theory behind the simulation optimization problem.

Example 6.1.1

This is the simple fuzzy differential equation

$$\dot{x} = \overline{A}x, \tag{6.2}$$

where $\overline{A} = (a_1/a_2/a_3) > 0$ and $x(0) = x_0 > 0$. As before, $\dot{x} = dx/dt$. The solution is

$$\overline{x}(t) = x_0 exp(\overline{A}t). \tag{6.3}$$

Hence,

$$x_1(t,0) = x_0 exp(a_1 t), \tag{6.4}$$

and

$$x_2(t,0) = x_0 exp(a_3 t). \tag{6.5}$$

Example 6.1.2

This is the "arms race" model in equations (3.1) and (3.2) in Chapter 3 adopted from [4]. The system of crisp differential equations are

$$\dot{x} = -ax + by + r, \tag{6.6}$$

and

$$\dot{y} = cx - dy + s. \tag{6.7}$$

We study this continuous system in more detail in Chapter 8. All the parameters $a, b, c, ..., s$ have uncertain values and must be estimated. Assume they are estimated by a group of experts (Chapter 3). Also assume that the initial values of x and y are known and so will be crisp and not fuzzy.

Now $\overline{a}$,...,$\overline{s}$ will all be triangular fuzzy numbers. Let $\overline{a}[0] = [a_1, a_2]$, $\overline{b}[0] = [b_1, b_2]$, etc. We know how to get $x(t,1)$ and $y(t,1)$: use the vertex values for all the triangular fuzzy numbers. But how shall we pick $a \in \overline{a}[0]$,...,$s \in \overline{s}[0]$ to obtain $x_i(t,0)$ $(y_i(t,0))$, $i = 1,2$? We can make a guess, but we have no proof that these values will do the job. For $x_1(t,0)$ make the slope $(\dot{x})$ as small as possible so use a_2,b_1,r_1. But $x(t)$ is also effected by $y(t)$ through by so also make $y(t)$ small. This guess, or the "intuitive" method, does not always work. So we need another method which will be explained in the next section.

6.2 Theory

In this section we will be using an example from [2]. Consider the second order linear ordinary differential equation

$$\ddot{x} + a\dot{x} + bx = c\cos(t), \tag{6.8}$$

where a, b and c are constants and $\ddot{x}$ $(\dot{x})$ is the second (first) time derivative of x. The only fuzzy parameters will be the initial conditions so let $\overline{x}(0) = \overline{x}_0 = (x_1/x_2/x_3)$ and the fuzzy initial value of $\dot{x} = \overline{v}_0 = (v_{01}/v_{02}/v_{03})$ both triangular fuzzy numbers. Using the fuzzy initial conditions the fuzzy solution is $\overline{x}(t)$. Let $\overline{x}(t)[\alpha] = [x_1(t,\alpha), x_2(t,\alpha)]$ for $t \geq 0$ and $0 \leq \alpha \leq 1$. We are interested in finding the graphs of $x_1(t,0)$ and $x_2(t,0)$.

Let us look at this problem another way. Define

$$\Omega(t) = \{x(t) | x_0 \in \overline{x}_0[0], v_0 \in \overline{v}_0[0]\}. \tag{6.9}$$

The graph of $\Omega(t)$ presents the region of uncertainty in the solutions due to the uncertainty in the initial conditions. To describe this region we find its boundary

$$boundary[\Omega(t)] = (x_1(t,0), x_2(t,0)). \tag{6.10}$$

This is why we want to graph the $x_i(t,0)$, $i = 1,2$.

Now we have a final way of analyzing the problem. We can get a closed form solution to equation (6.8) giving

$$x(t) = F(t; a, b, c, x_0, v_0). \tag{6.11}$$

We fuzzify this solution using the extension principle

$$\overline{x}(t) = F(t; a, b, c, \overline{x}_0, \overline{v}_0). \tag{6.12}$$

Assuming that f is continuous we know that [3]

$$x_1(t,0) = min\{F(t; a, b, c, x_0, v_0) | x_0 \in \overline{x}_0[0], v_0 \in \overline{v}_0[0]\}, \tag{6.13}$$

and

$$x_2(t,0) = max\{F(t; a, b, c, x_0, v_0) | x_0 \in \overline{x}_0[0], v_0 \in \overline{v}_0[0]\}. \tag{6.14}$$

Given values for a, b and c, for this simple problem, we may solve equations (6.13) and (6.14) for the value of x_0 and v_0 that produce the min and max. The results are that there are $0 < \delta_1 < \delta_2 < ...$ so that:

1. for $t \in [0, \delta_1]$

$$x_1(t,0) = F(t; a, b, c, x_1, v_{01}), \tag{6.15}$$

$$x_2(t,0) = F(t; a, b, c, x_3, v_{03}); \tag{6.16}$$

2. for $t \in [\delta_1, \delta_2]$

$$x_1(t,0) = F(t; a, b, c, x_3, v_{01}), \tag{6.17}$$

$$x_2(t,0) = F(t; a, b, c, x_1, v_{03}); \tag{6.18}$$

3. for $t \in [\delta_2, \delta_3]$

$$x_1(t,0) = F(t; a, b, c, x_3, v_{03}), \tag{6.19}$$

$$x_2(t,0) = F(t; a, b, c, x_1, v_{01}); \tag{6.20}$$

4. for $t \in [\delta_3, \delta_4]$

$$x_1(t,0) = F(t; a, b, c, x_1, v_{03}), \tag{6.21}$$

$$x_2(t,0) = F(t; a, b, c, x_3, v_{01}); \tag{6.22}$$

5. etc. same pattern.

This shows that, in general, you can not keep the same values for x_0 and v_0 for all t and obtain the boundary of $\Omega(t)$. You can always tell that you have lost the description of the boundary when the graphs of $x_1(t,0)$ and $x_2(t,0)$ cross and $x_1(t,0) > x_2(t,0)$.

The above results are for one second order linear differential equation fuzzifing only the initial conditions. However, we believe the same thing can happen for systems of (non)linear differential equations fuzzifing all the parameters.

Unfortunately, the above analysis can not be performed in the examples presented in Chapters 7 - 26 because: (1) either there is no known closed form solutions employing elementary functions; or (2) the closed form solutions are too complicated. So we will be unable to find the times δ_i to change the values of the parameters in their $\alpha = 0$ cut to get the boundary functions $x_i(t,0)$.

There is a solution to the problem of determining the graph of $\Omega(t)$. Now $\Omega(t)$ represents all the graphs of a crisp solution $x(t)$ as all the fuzzy parameters range through their intervals of uncertainty. Our simulation software allows us to simulate the system as you sweep parameter values across a range of values. Code for this is in Chapter 28. Of course, this "sweep" will have to be discrete. Let the uncertain parameters be a_i, $1 \leq i \leq k$. Assume they will all have fuzzy estimators $\overline{a}_i$, $1 \leq i \leq k$. These fuzzy estimators will all be triangular fuzzy numbers $\overline{a}_i = (a_{i1}/a_{i2}/a_{i3})$, $1 \leq i \leq n$. For a minimum approximation to $\Omega(t)$, the region of uncertainty, we will graph all the crisp solutions on the same graph for $a_i = a_{i1}, a_{i2}, a_{i3}$, $1 \leq i \leq k$. So there will be 3^k graphs on the same coordinate system producing an approximation to $x_i(t,0)$, $i = 1,2$. In the applications in Chapters 7 - 26 k can range from 2 to 11. Surely, if $n = 11$, then $3^{11} = 177,147$ which is too many graphs. We will need to reach another compromise. Our final compromise is that we will allow at most six fuzzy parameters in each application. The result is : (1) for two fuzzy parameters, there will be 9 graphs; (2) if $k = 3$ we have 27 graphs; (3) with four fuzzy parameters we get 81 graphs; (4) $k = 5$ gives 243 graphs; and (5) $k = 6$ we have 729 graphs on one coordinate system. See Chapter 28 if you want to use more than three values of a fuzzy parameter and/or more than six fuzzy variables. The ideal would be to use thousands of graphs producing a black tube of uncertainty for Ω.

Sometimes we will have only one fuzzy parameter. If we chose only three crisp values for this parameter in its $\alpha = 0$ cut we would have just three graphs which would be too few to approximate the band of uncertainty. So in this case we will pick around ten equally spaced crisp values for this parameter in its $\alpha = 0$ cut. We will make these special changes in choosing the values for the fuzzy parameters when necessary.

6.3 Summary

Consider a crisp continuous system whose evolution is determined by a system of ordinary differential equations. Some parameter values are not known precisely and must be estimated. Using our fuzzy estimators (Chapter 3) we get a system of fuzzy differential equations to solve. If $x(t)$ is a crisp solution we wish to study, then it becomes a fuzzy trajectory $\overline{x}(t)$ in the fuzzy system. We want to identify $\Omega(t)$, which is the collection of all crisp trajectories $x(t)$ obtained as all the uncertain parameters range through their $\alpha = 0$ cuts. $\Omega(t)$ gives the uncertainty band for all possible crisp solutions as a result of the uncertainty in parameter values. We find $\Omega(t)$ by determining its boundary $x_1(t, 0)$ and $x_2(t, 0)$. The simulation optimization problem is to find out how to pick the values of the parameters in their alpha zero cut to obtain the boundary of $\Omega(t)$. We have not solved this problem in general and this is surely a topic for future research.

6.4 References

1. J.J.Buckley: Simulating Fuzzy Systems, Springer, Heidelberg, Germany, 2005.

2. J.J.Buckley and T.Feuring: Fuzzy Initial Value Problem for Nth Order Linear Differential Equations, Fuzzy Sets and Systems, 121(2001)247-255.

3. J.J.Buckley and Y.Qu: On Using α-Cuts to Evaluate Fuzzy Equations, Fuzzy Sets and Systems, 38(1990)309-312.

4. M.Olinick: An Introduction to Mathematical Models in the Social and Life Sciences, Addison-Wesley, Reading, MA, 1978.

Chapter 7

Predator/Prey Models

7.1 Introduction

The predator/prey model was introduced in Chapter 4. However, in this chapter we will modify it as discussed below. Let $x(t)$ $(y(t))$ denote the number of fox (rabbits) in a certain ecosystem at time t. Assume that in this region the fox only eat rabbits and therefore the $x(t)$ are the predators and the $y(t)$ are the prey. If there are no rabbits the fox population may decline according to

$$\dot{x} = -ax, \; a > 0. \tag{7.1}$$

If rabbits are present we assume that the fox population will grow at a rate proportional to the product xy, or at rate bxy. Hence

$$\dot{x} = -ax + bxy, \; a > 0, \; b > 0, \tag{7.2}$$

is the differential equation governing the fox population. We use the notation $\dot{x} = dx/dt$.

If there are no foxes, assuming unlimited food supply, then $\dot{y} = dy$, for some constant $d > 0$. But this allows the rabbit population to grow indefinitely. Suppose there is some theoretical maximum number (y_m) of rabbits that can live in this ecosystem; or the carrying capacity is y_m. Then we write $\dot{y} = dy(1 - (y/y_m))$ [1]. When there are foxes present, the rabbit population will decline at rate cxy, $c > 0$. So

$$\dot{y} = -cxy + dy(1 - y/y_m)), \; c > 0, \; d > 0. \tag{7.3}$$

This is the differential equation describing the rabbit population and both differential equations (equations (7.2) and (7.3)) is the model for the system ([1],[2]). This is a system of nonlinear differential equations. The initial conditions are $x(0) = x_0$ and $y(0) = y_0$. If there is a value of t, say $t = t_z$, for which $x(t_z) < 1$ $(y(t_z) < 1)$, then $x(t) = 0$ $(y(t) = 0)$ for $t \geq t_z$. Since these

Parameter	Fuzzy/Crisp Value
a	$\overline{a} = (0.2/0.3/0.4)$
b	$\overline{b} = (0.04/0.05/0.06)$
c	$\overline{c} = (0.3/0.4/0.5)$
d	$\overline{d} = (5.5/6/6.5)$
y_m	100
x_0	5
y_0	15

Table 7.1: Fuzzy/Crisp Parameters in the Predator/Prey Model

are continuous models we could have $x(t) = 0.1$ for some value of t which is impossible since $x(t)$ needs to be $0, 1, 2, ..., y_m$.

7.2 Parameters

We need to estimate all of the parameters. We have no data on this ecosystem so we turn to experts (Chapter 3) on this region to help estimate a, b, c, d, y_m and x_0, y_0 for a certain time period. We obtained the following triangular fuzzy numbers shown in Table 7.1. Notice that y_m, x_0 and y_0 are assumed to be known precisely and therefore are not fuzzy. Using fuzzy parameters we get fuzzy differential equations and the solutions $\overline{x}(t)$ and $\overline{y}(t)$ are also fuzzy. The system now becomes a continuous fuzzy system whose trajectories are fuzzy so that any slice through a trajectory at some time t_z is a fuzzy number. We wish to estimate the band of uncertainty which is $\overline{x}(t)[0]$ and $\overline{y}(t)[0]$, or the $\alpha = 0$ cuts.

Next we need to choose the values of the parameters in their $\alpha = 0$ cut to approximate the outer boundary of the $\alpha = 0$ cut of the fuzzy trajectory for the foxes and the rabbits. As discussed in Chapter 6 if $\overline{\omega} = (\omega_1/\omega_2/\omega_3)$ is a fuzzy parameter we use $\omega = \omega_1$, ω_2 and ω_3. We do this for $\overline{a}$, $\overline{b}$, $\overline{c}$ and $\overline{d}$ giving $3^4 = 81$ curves for $x(t)$ and 81 graphs for $y(t)$ for a minimal approximation to $\overline{x}(t)[0]$ and $\overline{y}(t)[0]$. These are shown in Figures 7.2 and 7.3. The systems diagram for Simulink is in Figure 7.1.

In Figure 7.2 (7.3) the reader can determine, from consulting Section 1.5.4, $x(t)$ ($y(t)$) when we used for the fuzzy parameters: (1) all left end points of their $\alpha = 0$ cut; (2) all right end points of their $\alpha = 0$ cut; and (3) their $\alpha = 1$ values.

7.3 Simulation

Let us first explain the Simulink network in Figure 7.1. The top loop is for $x(t)$ and the lower loop is for $y(t)$. The input to "Integrator" in the top loop

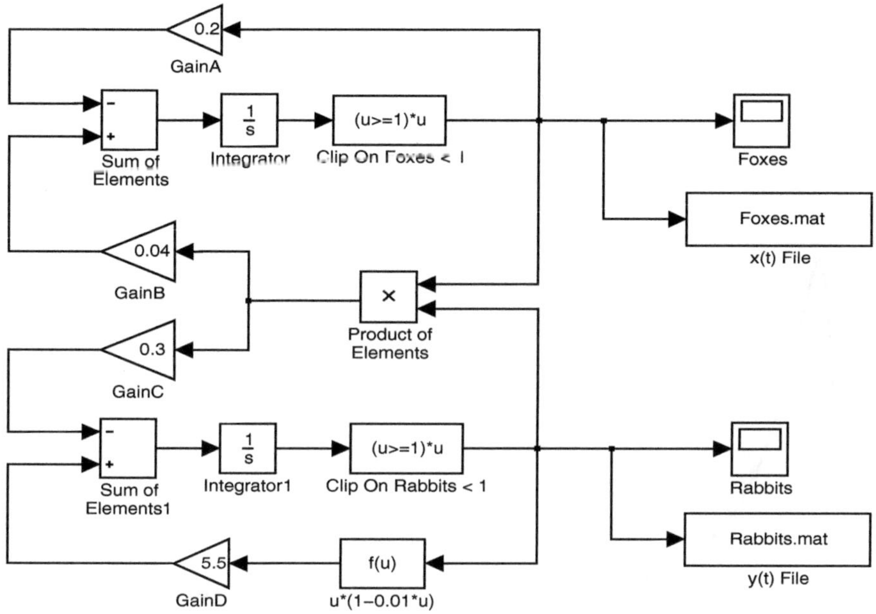

Figure 7.1: Simulink Diagram for the Predator/Prey Model

is $-ax + bxy$ which equals $\dot{x}$ and the output is $x(t)$. The initial value for $x(t)$ was set inside the "Integrator" block. "Clip on Foxes" sets $x(t)$ to zero when it becomes less than one. The function within the box, $(u >= 1) * u$, first interprets $u >= 1$ as true/false, yielding 1/0, and subsequently produces an output of $u/0$. "GainA" multiplies $x(t)$ by a for input to "Sum of Elements". $x(t)$ is also input to "Product of Elements" which computes $x(t)y(t)$, and then gets multiplied by b in "GainB" for input to "Sum of Elements". The input to "Integrator1" in the lower loop is $-cxy+dy(1-y/y_m)$ which equals $\dot{y}$ and the output is $y(t)$. The initial value for $y(t)$ was set inside "Integrator1". "Clip on Rabbits" makes $y(t)$ zero if $y(t) < 1$, as is done for "Clip on Foxes". $y(t)$ goes to "Product of Elements" and also to "f(u)". The block "f(u)" computes $y(1-y/y_m)$, whose output goes to "GainD" for multiplication by d, and its output goes to "Sum of Elements1". Also, the output from "Product of Elements", which is $x(t)y(t)$, gets multiplied by c for input to "Sum of Elements1". "Scope" produces the graph of $x(t)$ and "Scope1" is for the graph of $y(t)$. "Scope" graphs only one curve at a time so "Foxes.mat" ("Rabbits.mat") collects the data for all 81 graphs for graphing on one coordinate system. More details on how this is done is in Chapter 28. There are also other items to set/choose, like simulation time and type of numerical integrator, which

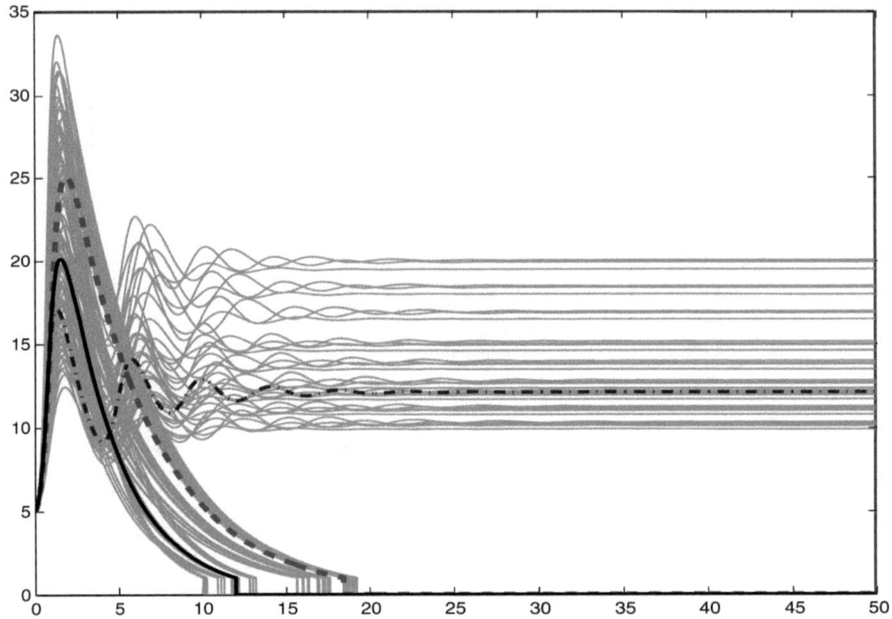

Figure 7.2: Fuzzy Trajectory for the Number of Foxes $\overline{x}(t)[0]$

we will not discuss here (see Chapter 28).

Let $\overline{x}(t)[0] = [x_1(t,0), x_2(t,0)]$. In Figure 7.2 connect the upper boundary with a smooth curve which is our approximation to $x_2(t,0)$ and connect the lower boundary with a smooth curve producing an approximation to $x_1(t,0)$. The horizontal axis is time and the vertical axis is the number of foxes. We get steady state (horizontal curves) with $\lim_{t\to\infty} x_2(t,0) \approx 20$ and $\lim_{t\to\infty} x_1(t,0) = 0$. The fox population can go to zero because the rabbit population can go to zero (Figure 7.3). The uncertainty band, the difference between the two curves in steady state, is approximately 20 foxes. But initially, the fox population can oscillate from a maximum of about 34 to a minimum of zero after starting out at 5. We actually obtain a steady state band of 10 to 20 if the fox population does not approach zero.

Let $\overline{y}(t)[0] = [y_1(t,0), y_2(t,0)]$. In Figure 7.3 connect the upper boundary with a smooth curve which is our approximation to $y_2(t,0)$ and connect the lower boundary with a smooth curve producing an approximation to $y_1(t,0)$. The horizontal axis is time and the vertical axis is the number of rabbits. We get steady state (horizontal curves) with $\lim_{t\to\infty} y_2(t,0) \approx 10$ and $\lim_{t\to\infty} y_1(t,0) = 0$. The rabbit population can go to zero. However, initially the rabbit population can go a high as around 54 after starting out at

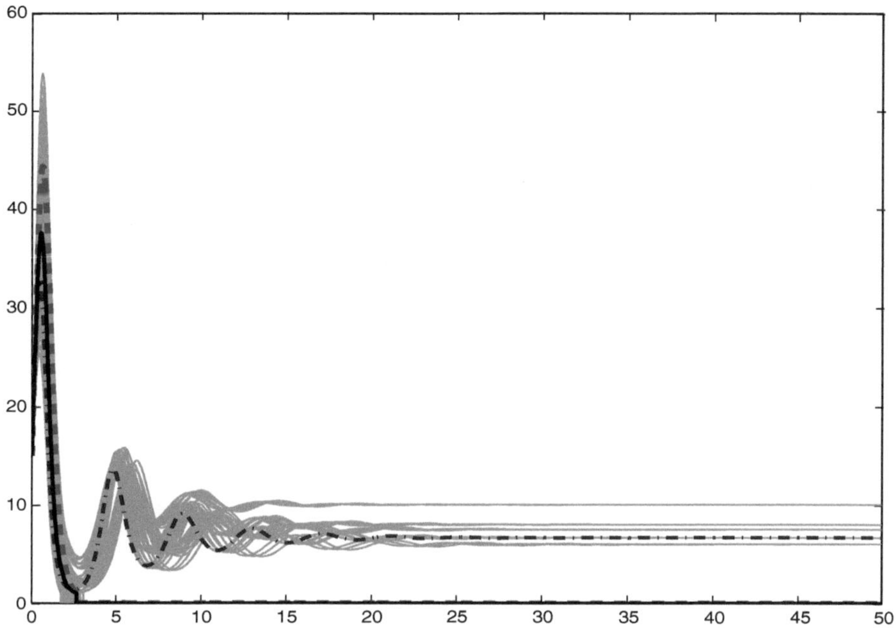

Figure 7.3: Fuzzy Trajectory for the Number of Rabbits $\overline{y}(t)[0]$

15. We also obtain a steady state uncertainty band of about 6 to 10 rabbits if the population does not go to zero.

The results show that there is a possibility that the population of foxes and rabbits will disappear. But there is a lot of uncertainty in the results due to the uncertainty in the values of the parameters in the model. If we can obtain more accurate estimates for these parameters (less fuzzy), then we can reduce the uncertainty in the results for $\overline{x}(t)$ and $\overline{y}(t)$.

If all the uncertain parameters are estimated from data (Chapter 3), instead of employing experts, then the base of the fuzzy estimator is a 99% confidence interval. Using all of these confidence intervals we get $\overline{x}(t)[0]$ and $\overline{y}(t)[0]$, so these bands of uncertainty will be like 99% confidence intervals for $x(t)$ and $y(t)$, respectively.

7.4 References

1. M.R.Spiegel: Applied Differential Equations, Third Edition, Prentice-Hall, Englewood Cliffs, NJ, 1981.

2. D.G.Zill: A First Course in Differential Equations, Brooks/Cole, Pacific Grove, CA, 1997.

Chapter 8

An Arm's Race Model

8.1 Introduction

The arms race model was briefly introduced in Chapter 3 and again in Chapter 6. The system of crisp linear differential equations is [1]

$$\dot{x} = -ax + by + r, \tag{8.1}$$

$$\dot{y} = cx - dy + s, \tag{8.2}$$

where a, b, c, d are all positive constants, r, s are positive, or negative, constants, subject to initial conditions $x(0) = x_0$, $y(0) = y_0$. Here x (y) represents the yearly rates of armament expenditures of nation A (B) in dollars. We use the notation $\dot{x} = dx/dt$ and $\dot{y} = dy/dt$. Time will be in years. These are linear differential equations so we could easily find the solutions. However, even using simple solutions it can be complicated to find the boundary of the band of uncertainty, or the boundary of $\overline{x}(t)[0]$) and $\overline{y}(t)[0]$, as shown in Section 6.2. So we omit expressions for the exact solutions and will use simulation instead.

We first need to look more closely at this model to understand the meaning of the constants a, b, c, d, r, s. The first assumption is that $\dot{x}$ ($\dot{y}$) is directly proportional to y (x). Hence $\dot{x} = by$ ($\dot{y} = cx$) for some positive constant b (c). The next assumption is that excessive expenditures places a "drag" on the economy, or $\dot{x}$ ($\dot{y}$) is directly and negatively proportional to x (y). That is $\dot{x} = -ax$ ($\dot{y} = -dy$) for some positive constant a (d). So now we have $\dot{x} = -ax + by$ and $\dot{y} = cx - dy$. Finally, increasing (decreasing) arms expenditures is not the only mutual stimulation on expenditures, but underlying grievance (good will) of each nation against the other have important implications on expenditures and are in the constants r and s in equations (8.1) and (8.2). More details are in [1]. We also need initial values for x and y. If there is a value of t, say $t = t_z$, for which $x(t_z) = 0$ ($y(t_z) = 0$), then we assume that $x(t) = 0$ ($y(t) = 0$) for $t \geq t_z$.

8.2 Parameters

We need to estimate all the parameters in the model. The data we have is old, from during and before the last great war. So we turn to a group of experts (Chapter 3) in this field to help estimate all the parameters. We obtained the triangular fuzzy numbers shown in Table 8.1. Notice that $x_0 = y_0 = 15$ are known precisely and will not be fuzzy. Using fuzzy parameters we get fuzzy differential equations and the solutions $\overline{x}(t)$ and $\overline{y}(t)$ are also fuzzy. The system now becomes a continuous fuzzy system whose trajectories are fuzzy so that any slice through a trajectory at some time t_z is a fuzzy number. We wish to estimate the band of uncertainty which is $\overline{x}(t)[0]$ and $\overline{y}(t)[0]$, or the $\alpha = 0$ cuts.

Parameter	Fuzzy/Crisp Value
a	$\overline{a} = (1.5/2/2.5)$
b	$\overline{b} = (0.5/1/1.5)$
c	$\overline{c} = (5.5/6/6.5)$
d	$\overline{d} = (1.5/2/2.5)$
r	$\overline{r} = (-5.5/-5/-4.5)$
s	$\overline{s} = (-12.5/-12/-11.5)$
x_0	15
y_0	15

Table 8.1: Fuzzy/Crisp Parameters in the Arm's Race Model for the First Simulation

Next we need to choose the values of the fuzzy parameters in their $\alpha = 0$ cut to approximate the outer boundary of the $\alpha = 0$ cut of the fuzzy trajectory for the yearly rates of armament expenditures for nation A (B). As discussed in Chapter 6 if $\overline{\omega} = (\omega_1/\omega_2/\omega_3)$ is a fuzzy parameter we use $\omega = \omega_1$, ω_2 and ω_3. We do this for $\overline{a}$, $\overline{b}$, $\overline{c}$, $\overline{d}$, $\overline{r}$ and $\overline{s}$ giving $3^6 = 729$ curves for $x(t)$ and 729 graphs for $y(t)$ for a minimal approximation to $\overline{x}(t)[0]$ and $\overline{y}(t)[0]$. These are shown in Figures 8.2 and 8.3. The systems diagram for Simulink is in Figure 8.1.

In Figure 8.2 (8.3) the reader can determine, from consulting Section 1.5.4, $x(t)$ ($y(t)$) when we used for the fuzzy parameters: (1) all left end points of their $\alpha = 0$ cut; (2) all right end points of their $\alpha = 0$ cut; and (3) their $\alpha = 1$ values.

8.3 First Simulation

Let us first describe how Simulink solves this problem using Figure 8.1. The top of the figure is for $x(t)$ and the bottom is for $y(t)$. The input to the top "Integrator" is $-ax + by + r$ which is $\dot{x}$ and the output is $x(t)$. The

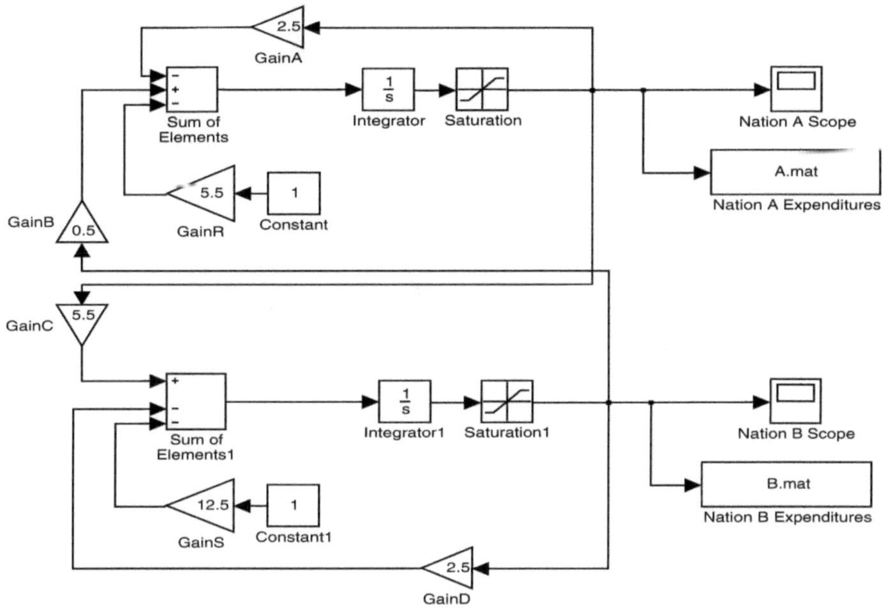

Figure 8.1: Simulink Diagram for Arm's Race Model

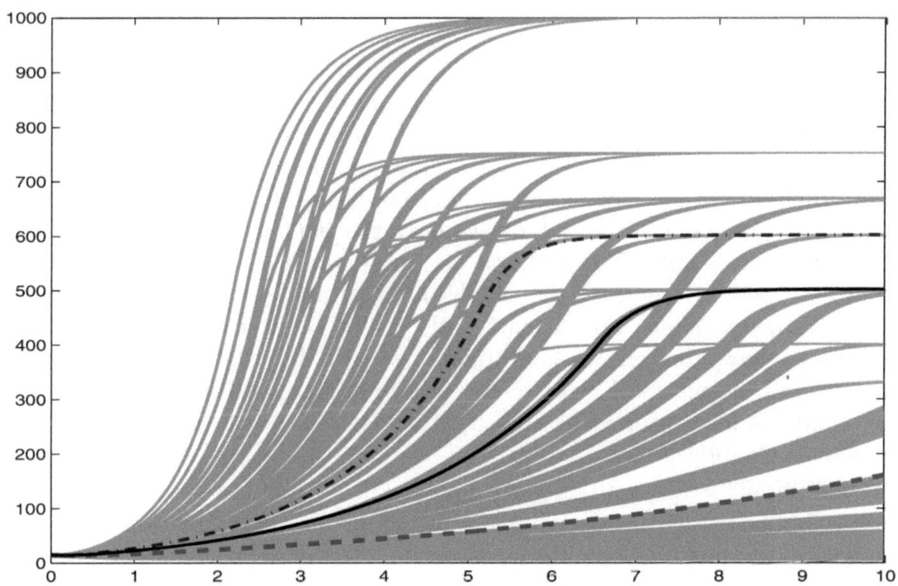

Figure 8.2: Fuzzy Trajectory for Nation A $\overline{x}(t)[0]$ in the Arm's Race Model in the First Simulation

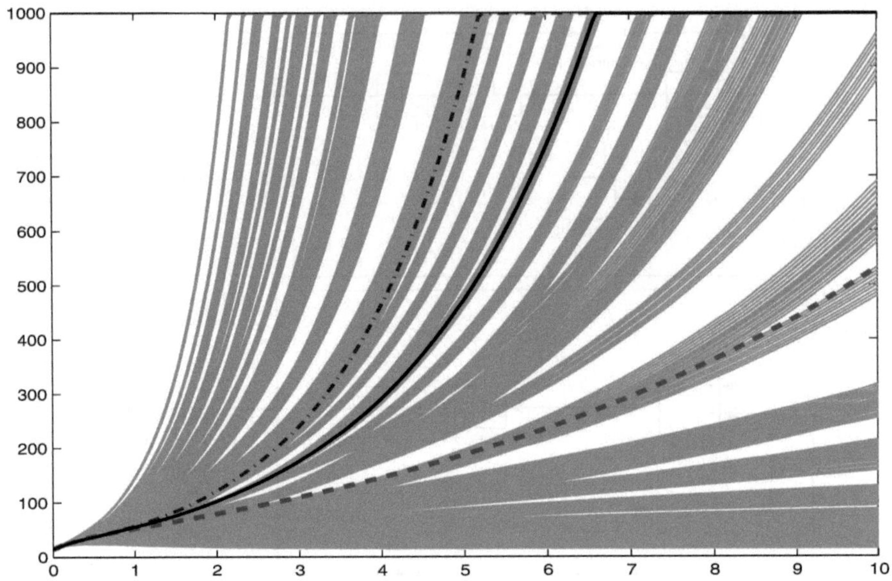

Figure 8.3: Fuzzy Trajectory for Nation B $\overline{y}(t)[0]$ in the Arm's Race Model in the First Simulation

"Saturation" just sets $x(t) = 0$ if $x(t)$ is negative (negative expenditures are not allowed) and it sets $x(t) = 1000$ if $x(t) > 1000$. We cut the simulation off at 1000 because we feel that the system is "out of control" when expenditures exceed 1000. The top loop multiplies $x(t)$ by a in "GainA" for input to "Sum of Elements". The second input to "Sum of Elements" is from $y(t)$ after multiplying by b in "GainB". The final input to "Sum of Elements" is the constant r in "GainR". Similarly we have the bottom half of the figure for $y(t)$. The initial values for $x(t)$ and $y(t)$ are defined inside the "Integrators" and the "Scopes" produce the graphs for $x(t)$ and $y(t)$ for each choice for the parameter values. So "A.mat" and "B.mat" collect the data for all 729 graphs for graphing on one coordinate system. More details on how this is done is in Chapter 28. There are also other items to set/choose, like the type of numerical integrator and the Max/Min step size, which we will not discuss here (see Chapter 28).

Let $\overline{x}(t)[0] = [x_1(t,0), x_2(t,0)]$ for nation A. In Figure 8.2 connect the upper boundary with a smooth curve which is our approximation to $x_2(t,0)$ and connect the lower boundary with a smooth curve producing an approximation to $x_1(t,0)$. The horizontal axis is time in years and the vertical axis is \$. We see that $x_2(t,0)$ can grow to exceed 1000 in about 5 years whereas $x_1(t,0)$ may approach zero which leads to disarmament.

Let $\overline{y}(t)[0] = [y_1(t,0), y_2(t,0)]$. In Figure 8.3 connect the upper boundary with a smooth curve which is our approximation to $y_2(t,0)$ and connect the lower boundary with a smooth curve producing an approximation to $y_1(t,0)$. The horizontal axis is time and the vertical axis is \$. We see that $y_2(t,0)$ can grow to exceed 1000 within 2 years whereas $y_1(t,0)$ may approach zero which leads to disarmament.

In both figures the spread between both graphs shows the uncertainty due to the uncertainty in the input parameters. In this example there is tremendous uncertainty in the results. The lower graphs $x_1(t,0)$ and $y_1(t,0)$ both approach zero, which leads to disarmament, in one to two years. The upper graphs keep increasing and eventually go off the top of the graph paper. The uncertainty is too great going from disarmament to "run away" expenditures on armament. There is a possibility of almost everything in between. This result may come from too much fuzziness in the fuzzy estimators. So we would now see if we could obtain less uncertainty (fuzziness) in the expert's estimation of the parameters and solve the problem again.

If all the uncertain parameters are estimated from data (Chapter 3), instead of using expert opinion, then the base of the fuzzy estimator is a 99% confidence interval. Using all of these confidence intervals we get $\overline{x}(t)[0]$ and $\overline{y}(t)[0]$, so these bands of uncertainty will be like 99% confidence intervals for $x(t)$ and $y(t)$, respectively.

8.4 Second Simulation

Back to the "experts" to estimate the uncertain parameters. After much discussion we obtain less fuzzy estimates shown in Table 8.2. With this new data we simulate the system again with results in Figures 8.4 and 8.5.

Parameter	Fuzzy/Crisp Value
a	$\overline{a} = (1.8/2/2.2)$
b	$\overline{b} = (0.8/1/1.2)$
c	$\overline{c} = (5.8/6/6.2)$
d	$\overline{d} = (1.8/2/2.2)$
r	$\overline{r} = (-5.2/-5/-4.8)$
s	$\overline{s} = (-12.2/-12/-11.8)$
x_0	15
y_0	15

Table 8.2: Fuzzy/Crisp Parameters in the Arm's Race Model for the Second Simulation

There is not much change from Figure 8.2 to Figure 8.4 and from Figure 8.3 to Figure 8.5. Almost anything is possible from "run away" expenditures on armament to disarmament. The experts do not believe they can further

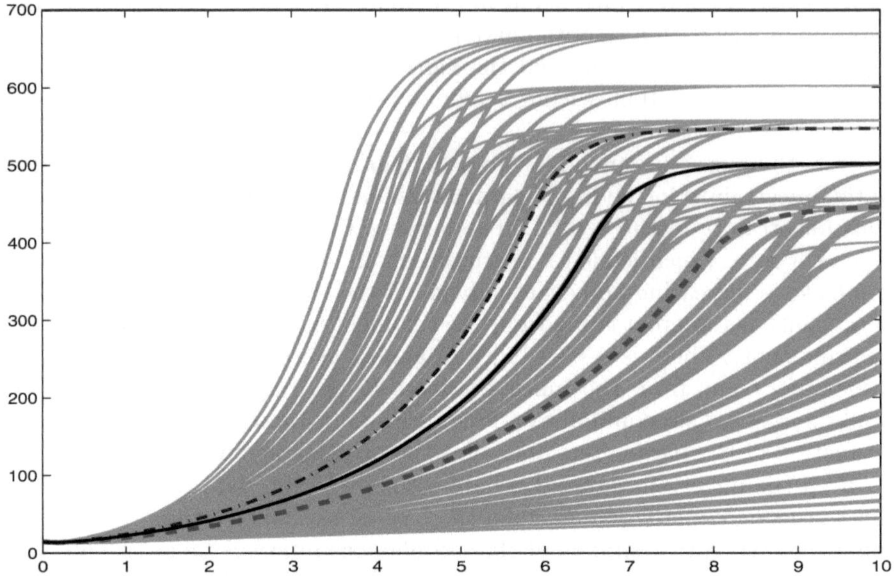

Figure 8.4: Fuzzy Trajectory for Nation A $\overline{x}(t)[0]$ in the Arm's Race Model in the Second Simulation

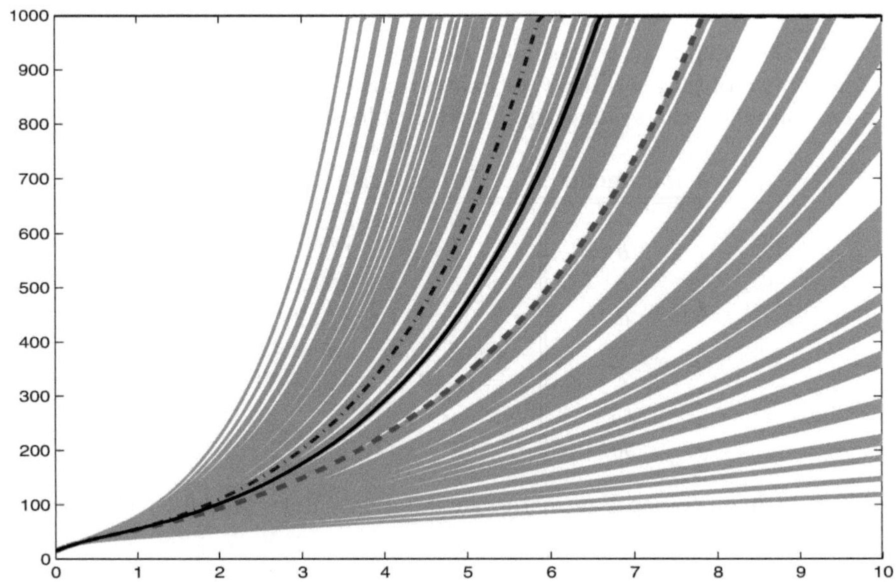

Figure 8.5: Fuzzy Trajectory for Nation B $\overline{y}(t)[0]$ in the Arm's Race Model in the Second Simulation

reduce the fuzziness in their estimates of the uncertain parameters. So, we can not make any predictions except that almost anything is possible.

8.5 References

1. M.Olinick: An Introduction to Mathematical Models in the Social and Life Sciences, Addison-Wesley, Reading, MA, 1978.

Chapter 9

Bungee Jumping

9.1 Introduction

This application was adapted from a tutorial/demo of Simulink [1]. We are to design a bungee jump for a amusement park in the US. A "rider" (customer) jumps from 240 feet above the ground. We will measure the positive x direction downward, from the point of the jump, to the ground level. The length of the unstretched bungee cord is 90 feet. Our coordinate system will be positive downward with $x = -90$ feet at the jump point, $x = 0$ where the bungee cord begins to stretch, and $x = 150$ feet at the ground. Initial conditions are then $x(0) = -90$ and initial velocity zero. The nonlinear differential equation of motion is

$$m(d^2x/dt^2) = mg - F(x) - R(v), \tag{9.1}$$

where m = mass, mg = weight of rider in pounds with $g = 32feet/sec^2$, $F(x)$ = the force on the jumper exerted by the bungee cord and $R(v)$ = is air resistance with v = velocity. We first have

$$F(x) = kx,\ x \geq 0,\ and\ F(x) = 0,\ x < 0, \tag{9.2}$$

where k is the spring constant for the elastic bungee cord. We will model air resistance simply as a constant times velocity

$$R(v) = cv, \tag{9.3}$$

where $v = \dot{x} = dx/dt$. We wish to find $x = x(t)$. Then for graphing we graph $150 - x(t)$.

9.2 Parameters

The weight of a customer and the constant c in air resistance are somewhat uncertain but for the bungee cord we plan to use, the constant k is known

Parameter	Fuzzy/Crisp Value
μ	$\overline{\mu} = (120/170/220)$
c	$\overline{c} = (1.0/1.2/1.4)$
k	$k = 2.5$

Table 9.1: Fuzzy/Crisp Parameters in the Bungee Jumping Model

to be $k = 2.5$ pounds per foot. We will model a customer's weight using a fuzzy estimator of the mean of a normal distribution (Section 3.6) and a fuzzy estimator for c will be from expert opinion (Section 3.2).

The customers that we are allowing to ride the bungee jump will be between 18 and 55 years old and usually male. We will design this ride only for males in this age bracket. We can get national data on the weight of this group of males and estimate the mean. Thus we can construct confidence intervals and obtain a fuzzy estimator $\overline{\mu}$ for μ as discussed in Section 3.6. This will be used for the fuzzy mass (weight divided by g) in the differential equation. For simplicity we will use a triangular fuzzy number for $\overline{\mu}$ instead of the triangular shaped fuzzy number shown in Figure 3.6.

The customers can produce different air resistance. It depends on what they are wearing (how loose the clothes) and their shape as they fall (legs together/apart, ...). Expert opinion is used to estimate c giving fuzzy number $\overline{c}$ as our estimator. Both fuzzy estimators are shown in Table 9.1. Using fuzzy parameters we get a fuzzy differential equation and the solution $\overline{x}(t)$ is also fuzzy. The system now becomes a continuous fuzzy system whose trajectories are fuzzy so that any slice through a trajectory at some time t_z is a fuzzy number. We wish to estimate the band of uncertainty which is $\overline{x}(t)[0]$, or the $\alpha = 0$ cut.

Next we need to choose the values of the parameters in their $\alpha = 0$ cut to approximate the outer boundary of the $\alpha = 0$ cut of the fuzzy trajectory. As discussed in Chapter 6 if $\overline{\omega} = (\omega_1/\omega_2/\omega_3)$ is a fuzzy parameter we use $\omega = \omega_1$, ω_2 and ω_3. We do this for $\overline{\mu}$ and $\overline{c}$ giving $3^2 = 9$ curves for $x(t)$ for a minimal approximation to $\overline{x}(t)[0]$. These are shown in Figure 9.2. The systems diagram for Simulink is in Figure 9.1.

In Figure 9.2 the reader can determine, from consulting Section 1.5.4, $x(t)$ when we used for the fuzzy parameters: (1) all left end points of their $\alpha = 0$ cut; (2) all right end points of their $\alpha = 0$ cut; and (3) their $\alpha = 1$ values.

9.3 First Simulation

Let us first describe how Simulink solves this problem using Figure 9.1. The output from the "Sum of Elements" is $mg - F(x) - R(v)$. "Gain" multiplies this by $1/m$ so the input to "Integrator" is $g - F(x)/m - R(v)/m$ which

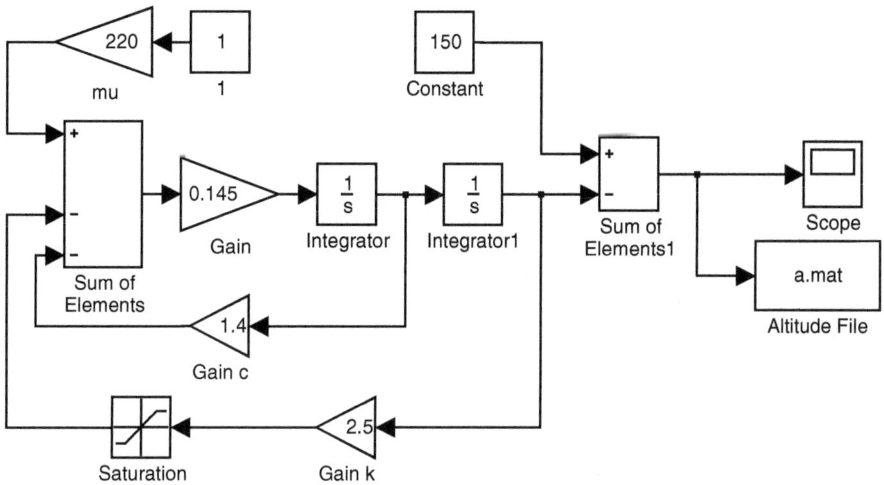

Figure 9.1: Simulink Diagram for the Bungee Jumping Example

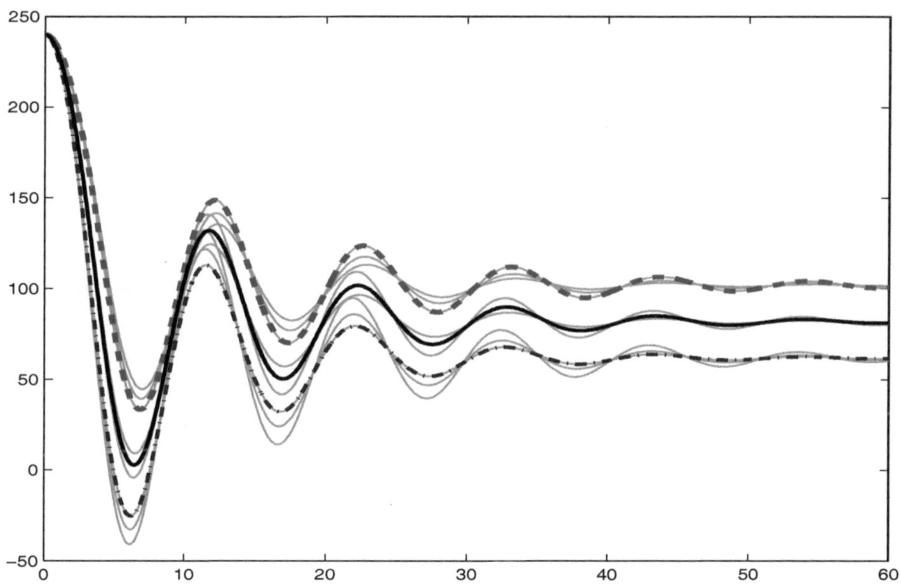

Figure 9.2: Fuzzy Trajectory for Altitude $\overline{x}(t)[0]$ in the Bungee Jumping Example: First Simulation

equals d^2x/dt^2. The output is $v = dx/dt$, which is the input to "Integrator1", whose output is $x(t)$. "Gain c" multiplies v by c for input to "Sum of Elements". "Gain k" multiplies x by k, then "Saturation" outputs zero when its input is negative, for input to "Sum of Elements". The constant "mu" is for the weight of the person jumping for input to "Sum of Elements". The initial values for v and x are set inside the "Integrators". Finally, "Sum of Elements1" computes $150 - x$ for "Scope" which gets the graph. "Scope" does a graph for only one value of μ and only one value for c. The "Altitude File" collects all the data for the nine graphs so they can be put into one coordinate system. More details on how this is done is in Chapter 28. There are other things to do, like choose the numerical integrator, choose the relative/absolute tolerance in the simulator, etc., but we will not discuss these here (see Chapter 28).

Let $\overline{x}(t)[0] = [x_1(t,0), x_2(t,0)]$. In Figure 9.2 connect the upper boundary with a smooth curve which is our approximation to $x_2(t,0)$ and connect the lower boundary with a smooth curve producing an approximation to $x_1(t,0)$. The horizontal axis is time and the vertical axis is distance above the ground. The spread between the graphs shows the uncertainty in the results due to the uncertainty in the input parameters. We see that there is a strong possibility that a customer may hit the ground, which is absolutely unacceptable. Therefore, we need to use another bungee cord with a higher value for k. We try again using $k = 4$, $k = 6$,... until the is no possibility of a jumper hitting the ground.

If all the uncertain parameters are estimated from data (Chapter 3), and not using expert opinion for $\overline{c}$, then the base of each fuzzy estimator is a 99% confidence interval. Using all of these confidence intervals we get $\overline{x}(t)[0]$, so this band of uncertainty will be like 99% confidence intervals for $x(t)$.

9.4 Second Simulation

We change k to 4 and run the simulation again. There are no other changes. The results are shown in Figure 9.3. Now there is no possibility that a person will hit the ground. We also want there to be the possibility that a rider will get close to the ground, making it a thrilling ride, and this is true as seen in Figure 9.3. Hence, we go with this design for the bungee jump.

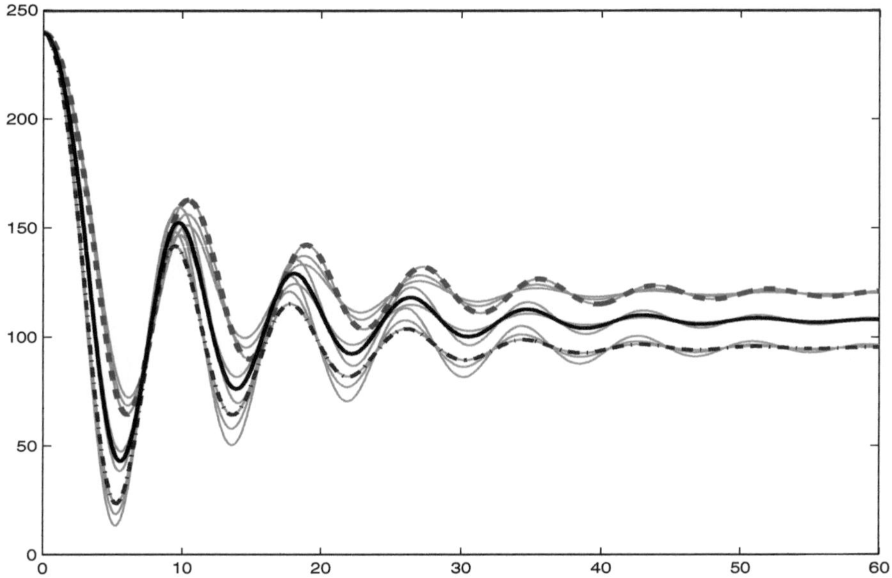

Figure 9.3: Fuzzy Trajectory for Altitude $\overline{x}(t)[0]$ in the Bungee Jumping Example: Second Simulation

9.5 Reference

1. www.mathworks.com

Chapter 10

Spread of Infectious Disease Model

10.1 Introduction

This application was adapted from examples in [1] and [2]. We have a total population of N people and a certain contagious disease infecting this population. Let us call this disease $\mathcal{D}$. This population is split up at any time t into three groups: (1) $x(t)$ = those uninfected with $\mathcal{D}$ but may become infected; (2) $y(t)$ = those who are presently infected with $\mathcal{D}$ and can spread the disease; and (3) $z(t)$ = those who had the disease $\mathcal{D}$ and are dead, recovered and immune, or isolated and can not spread the infection. We always have $N = x(t) + y(t) + z(t)$ for $t \geq 0$.

Our first assumption is that the rate of transfer from x into y is directly proportional to xy. Or, $\dot{x} = -kxy$ for some positive constant (the infection rate) k. This parameter needs to be estimated. The rate of transfer into y comes from x but the rate of transfer out of y goes to z. The rate out of y is assumed to be proportional to y. The differential equation is $\dot{y} = kxy - cy$ for some positive constant (the removal rate) c. This constant will also need to be estimated. If we differentiate $N = x + y + z$ and solve for dz/dt we obtain $dz/dt = cy$. But we do not need this third differential equation because we may always find z from $z = N - x - y$. Therefore, the system of nonlinear differential equations to solve is

$$\dot{x} = -kxy, \tag{10.1}$$

$$\dot{y} = kxy - cy, \tag{10.2}$$

$$z = N - x - y, \tag{10.3}$$

with initial conditions $x(0) = x_0$, $y(0) = y_0$, $z(0) = z_0 = 0$, $x_0 + y_0 = N$. The notation is $dx/dt = \dot{x}$ and $dy/dt = \dot{y}$.

Parameter	Fuzzy/Crisp Value
k	$\overline{k} = (0.003/0.005/0.007)$
c	$\overline{c} = (0.6/0.9/1.2)$
p	$\overline{p} = (0.02/0.05/0.08)$
y_0	$\overline{y}_0 = (20/50/80)$
x_0	$\overline{x}_0 = (920/950/980)$
z_0	$z_0 = 0$
N	$N = 1000$

Table 10.1: Fuzzy/Crisp Parameters in the Infectious Disease Model

10.2 Parameters

The constants k and c depend on the type of disease, the season (winter or summer), whether or not the population has been vaccinated against $\mathcal{D}$, etc. Experts in infectious diseases are asked to estimate k and c (Chapter 3). The resulting triangular fuzzy numbers for k and c are in Table 10.1. Next we need to estimate the initial value for y since we assumed that $z(0) = 0$ and then $x_0 = N - y_0$. Let $p =$ be the initial proportion of the population infected with $\mathcal{D}$. The value of p is not known precisely and needs to be estimated. The fuzzy estimator for p, see Section 3.5, is $\overline{p}$. Assume we obtained the triangular fuzzy number $\overline{p}$ shown in Table 10.1. There are only three basic fuzzy parameters in Table 10.1. They are $\overline{k}$, $\overline{c}$ and $\overline{p}$. Then $\overline{y}_0 = N\overline{p}$ and $\overline{x}_0 = N - \overline{y}_0$. For the rest of this chapter assume that $N = 1000$.

Using fuzzy parameters we get fuzzy differential equations and the solutions $\overline{x}(t)$, $\overline{y}(t)$ and $\overline{z}(t)$ are also fuzzy. The system now becomes a continuous fuzzy system whose trajectories are fuzzy so that any slice through a trajectory at some time t_z is a fuzzy number. We wish to estimate the bands of uncertainty which are $\overline{x}(t)[0]$, $\overline{y}(t)[0]$ and $\overline{z}(t)[0]$, or the $\alpha = 0$ cuts.

Next we need to choose the values of the parameters in their $\alpha = 0$ cut to approximate the outer boundary of the $\alpha = 0$ cut of the fuzzy trajectory for $\overline{x}(t)$, $\overline{y}(t)$ and $\overline{z}(t)$. As discussed in Chapter 6 if $\overline{\omega} = (\omega_1/\omega_2/\omega_3)$ is a fuzzy parameter we use $\omega = \omega_1$, ω_2 and ω_3. We do this for $\overline{k}$, $\overline{c}$ and $\overline{p}$ giving $3^3 = 27$ curves for $x(t)$, 27 graphs for $y(t)$, and 27 for $z(t)$, for a minimal approximation to $\overline{x}(t)[0]$, $\overline{y}(t)[0]$ and $\overline{z}(t)[0]$, respectively. These are shown in Figures 10.2 - 10.4. For example, if: (1) we pick $p = 0.02$, then $y_0 = 20$ and $x_0 = 980$; (2) for $p = 0.05$, then $y_0 = 50$ and $x_0 = 950$; and (3) $p = 0.08$ implies $y_0 = 80$, $x_0 = 920$. In all cases $z_0 = 0$ and $x_0 + y_0 = N$. The systems diagram for Simulink is in Figure 10.1.

In Figure 10.2 (10.3,10.4) the reader can determine, from consulting Section 1.5.4, $x(t)$ ($y(t)$,$z(t)$) when we used for the fuzzy parameters: (1) all left end points of their $\alpha = 0$ cut; (2) all right end points of their $\alpha = 0$ cut; and (3) their $\alpha = 1$ values.

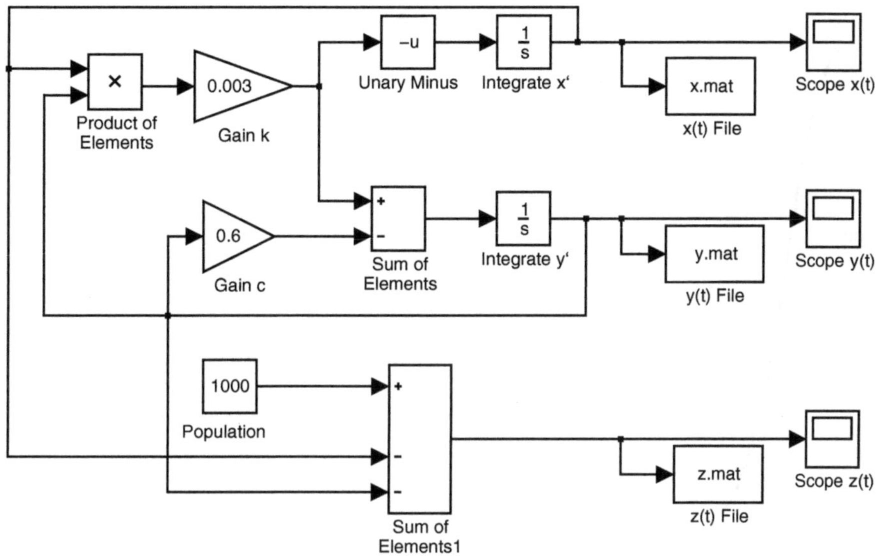

Figure 10.1: Simulink Diagram for the Infectious Disease Example

10.3 Simulation

We first discuss Figure 10.1. The top "Scope $x(t)$" is for $x(t)$, the middle "Scope $y(t)$" graphs $y(t)$, and the bottom "Scope $z(t)$" shows $z(t)$. As usual, a "Scope" only does one graph at a time so "$x(t)$ File" ("$y(t)$ File", "$z(t)$ File") collects the data for all the graphs so that we can display them on one coordinate system. More details on this is in Chapter 28. The input to "Integrate x' " is $dx/dt = -kxy$ with output $x(t)$. The input to "Integrate y' " is $dy/dt = kxy - cy$ with output $y(t)$. The bottom of the figure computes $z(t) = 1000 - x(t) - y(t)$ for "Scope $z(t)$". Now "Gain k" multiplies by k and then "Unary Minus" changes the sign, and "Gain c" multiplies by c. "Product of Elements" just multiplies the inputs and "Sum of Elements" adds/subtracts the inputs. Initial values for x and y are set inside the "Integrators". Other items, like simulation times and type of numerical integration to use, needed to run the simulation are not discussed here (see Chapter 28).

Let $\overline{x}(t)[0] = [x_1(t,0), x_2(t,0)]$. In Figure 10.2 connect the upper boundary with a smooth curve which is our approximation to $x_2(t,0)$ and connect the lower boundary with a smooth curve producing an approximation to $x_1(t,0)$. The horizontal axis is time and the vertical axis is the number uninfected with the disease. We get steady state (horizontal curves) with $\lim_{t\to\infty} x_2(t,0) \approx 100$ and $\lim_{t\to\infty} x_1(t,0) = 0$. The uncertainty band, the

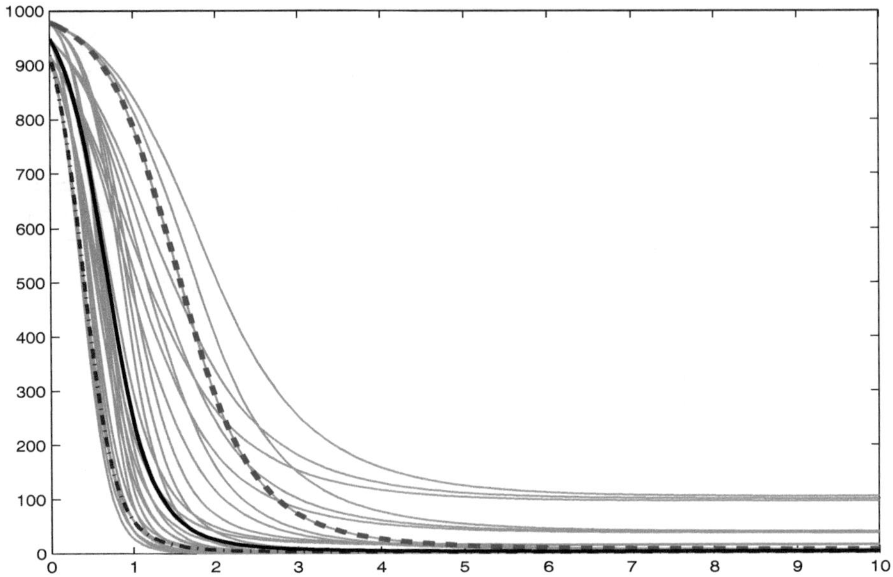

Figure 10.2: Fuzzy Trajectory for Uninfected $\overline{x}(t)[0]$ in the Infectious Disease Example

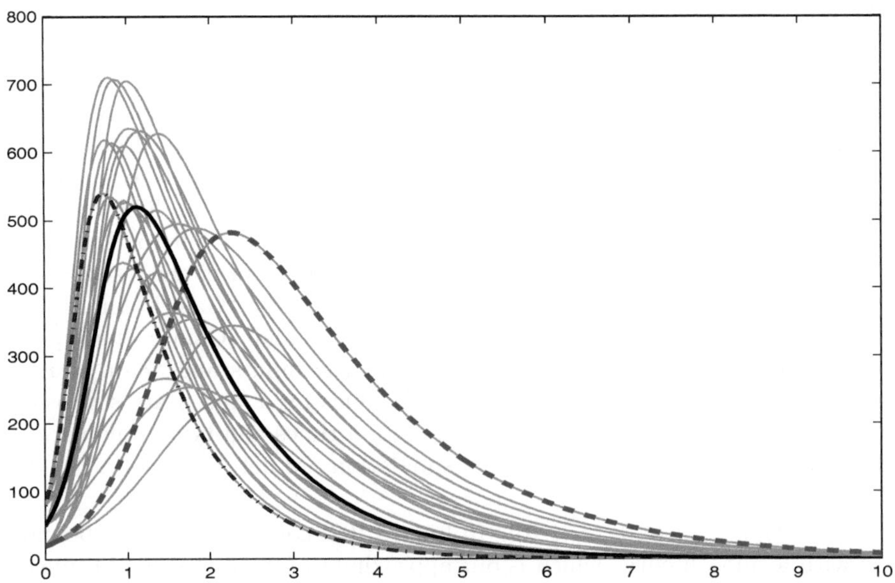

Figure 10.3: Fuzzy Trajectory for Infected $\overline{y}(t)[0]$ in the Infectious Disease Example

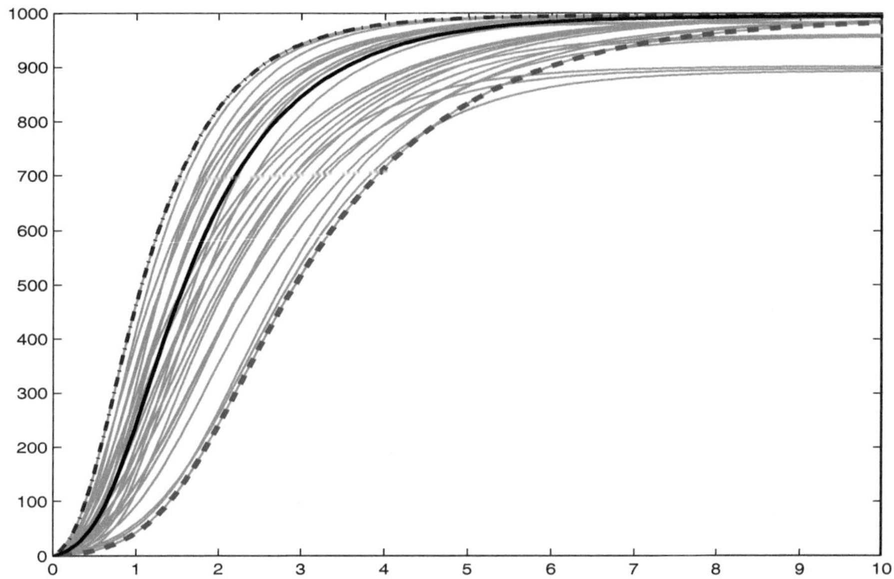

Figure 10.4: Fuzzy Trajectory for Previously Infected $\overline{z}(t)[0]$ in the Infectious Disease Example

difference between the two curves in steady state, is approximately 100 people. This uncertainty is a result of the uncertainty in k, c and p.

Let $\overline{y}(t)[0] = [y_1(t,0), y_2(t,0)]$. In Figure 10.3 connect the upper boundary with a smooth curve which is our approximation to $y_2(t,0)$ and connect the lower boundary with a smooth curve producing an approximation to $y_1(t,0)$. The horizontal axis is time and the vertical axis is the number of infected people with this disease. We get steady state (horizontal curves) with $\lim_{t\to\infty} y_2(t,0) \approx 0$ and $\lim_{t\to\infty} y_1(t,0) = 0$. The number infected goes to zero. As time grows there is no uncertainty in these results. The uncertainty band goes to zero. However, initially the number infected can go as high as around 710.

Let $\overline{z}(t)[0] = [z_1(t,0), z_2(t,0)]$. In Figure 10.4 connect the upper boundary with a smooth curve which is our approximation to $z_2(t,0)$ and connect the lower boundary with a smooth curve producing an approximation to $z_1(t,0)$. The horizontal axis is time and the vertical axis is the number in group $z(t)$. We get steady state (horizontal curves) with $\lim_{t\to\infty} z_2(t,0) = 1000$ and $\lim_{t\to\infty} x_1(t,0) \approx 900$. The uncertainty band, the difference between the two curves in steady state, is approximately 100 people. This uncertainty is a result of the uncertainty in $x(t)$.

The uncertainty bands in the results are not considered excessive, so for now we will consider this study finished.

If all the uncertain parameters are estimated from data (Chapter 3), and not using expert opinion, then the base of the fuzzy estimator is a 99% confidence interval. Using all of these confidence intervals we get $\overline{x}(t)[0]$,$\overline{y}(t)[0]$ and $\overline{z}(t)[0]$, so these bands of uncertainty will be like 99% confidence intervals for $x(t)$,$y(t)$ and $z(t)$, respectively.

10.4 References

1. M.Olinick: An Introduction to Mathematical Models in the Social and Life Sciences, Addison-Wesley, Reading, MA, 1978.

2. M.R.Spiegel: Applied Differential Equations, Third Edition, Prentice-Hall, Englewood Cliffs, NJ, 1981.

Chapter 11

Planetary Motion

11.1 Introduction

This application was adapted from a discussion on planetary motion in [1]. From Newton's Law of Universal Gravitation, two objects of mass M_1 and M_2 a distance r apart, are attracted towards each other with a force F

$$F = (GM_1M_2)/r^2, \tag{11.1}$$

where G is the gravitational constant. Let us consider the motion of the earth around the sun, neglecting the effects of the other planets and moons.

Assume an xy coordinate system with the sun, as a point mass, at the origin and the earth, also as a point mass, at a position with coordinates (x, y). See Figure 11.1. Using $F = ma$ and after some algebra we obtain the nonlinear system of differential equations of motion

$$\ddot{x} = -kx/(x^2 + y^2)^{3/2}, \tag{11.2}$$

$$\ddot{y} = -ky/(x^2 + y^2)^{3/2}, \tag{11.3}$$

where $k = GM_s$, $M_s =$ the mass of the sun and $\ddot{x}$ $(\ddot{y})$ is the second time derivative of x (y). For initial conditions we place the earth at position $x = a > 0$ on the $x-$axis moving in the positive y direction with velocity v_0. Therefore, $x(0) = a$, $y(0) = 0$, $\dot{x}(0) = 0$, $\dot{y}(0) = v_0$ at $t = 0$. Let $x(t)$ and $y(t)$ be solutions to the above system of differential equations.

11.2 Parameters

We will measure mass in kilograms, time in seconds, force in newtons and distance in meters (or $km =$ kilometers). Then [1] G is $6.6720(10)^{-11}$ Nm^2/kg^2. However, it is the mass of the sun that is fuzzy. Figures for M_s give around

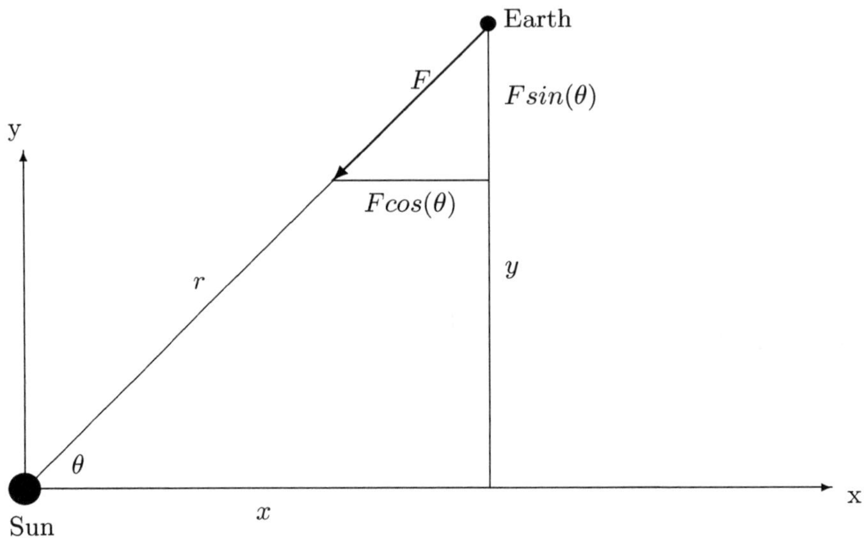

Figure 11.1: Diagram for the Earth's Orbit

$1.99(10)^{30}$ kg. But through nuclear fusion it is destroying 600 million metric tons of hydrogen each second [2]. Also, estimates of M_s are all through indirect calculations since we can not put it on a scale. For the mass of the sun we will take the fuzzy estimator $\overline{M}_s = (1.989/1.991/1.993)(10)^{30}$ kg. We will assume that the value of G is crisp and given above.

Next we need to find the value of a and v_0. We know that the orbit of the earth around the sun is an ellipse with the sun at a focal point. We put the sun at the origin so the second focal point will be on the negative x axis. The closest the earth comes to the sun will be at the point $(a, 0)$, called the perihelion, is approximately $149.577(10)^6$ km [1]. We will use a triangular fuzzy number estimator $\overline{a} = (140/150/160)10^6$ km for the perihelion of earth. We may derive an expression for v_0 as [1]

$$v_0 = \sqrt{GM_s(1+\epsilon)/a}, \tag{11.4}$$

where ϵ = eccentricity of earth's orbit known to be 0.0167. Using the known crisp/fuzzy values in this equation we may produce fuzzy values for v_0 km/sec. The values for all the parameters in this model are shown in Table 11.1. We will assume that the triangular fuzzy numbers for $\overline{M}_s$ and $\overline{a}$ were all obtained from experts (Chapter 3) but $\overline{v}_0$ is computed from equation (11.4) and $\overline{k}$ was computed from $G\overline{M}_s$.

Let $z(t) = (x(t), y(t))$. We are now interested in the graph of $z(t)$ showing the path of the earth around the sun. Using fuzzy parameters the system becomes a continuous fuzzy system whose trajectories $\overline{z}(t)$ are fuzzy so that any slice through a trajectory at some time t_z is a fuzzy number. We wish to estimate the band of uncertainty which is $\overline{z}(t)[0]$, or the $\alpha = 0$ cuts.

Parameter	Fuzzy/Crisp Value
G	$6.6720(10)^{-11}$
$\overline{M}_s$	$(1.989/1.991/1.993)(10)^{30}$
$\overline{v}_0$	From Equation (11.4)
$\overline{a}$	$(140/150/160)(10)^6$
$\overline{k}$	$= G\overline{M}_s$

Table 11.1: Fuzzy/Crisp Parameters for the Orbit of the Earth Around the Sun

Next we need to choose the values of the fuzzy parameters in their $\alpha = 0$ cut to approximate the outer boundary of the $\alpha = 0$ cut of the fuzzy trajectory for the orbit of the earth around the sun. As discussed in Chapter 6 if $\overline{\omega} = (\omega_1/\omega_2/\omega_3)$ is a fuzzy parameter we use $\omega = \omega_1$, ω_2 and ω_3. We have basically only two fuzzy sets in the system so we do this for $\overline{M}_s$ and $\overline{a}$ giving $3^2 = 9$ curves for $z(t)$ for a minimal approximation to $\overline{z}(t)[0]$. Once we choose values for M_s and a we calculate values for v_0 and k as discussed above. The fuzzy orbit is shown in Figure 11.3. The systems diagram for Simulink is in Figure 11.2.

In Figure 11.3 the reader can determine, from consulting Section 1.5.4, $z(t)$ when we used for the fuzzy parameters: (1) all left end points of their $\alpha = 0$ cut; (2) all right end points of their $\alpha = 0$ cut; and (3) their $\alpha = 1$ values.

11.3 Simulation

Let us first discuss Figure 11.2. The upper loop is for $x(t)$ and the bottom loop is for $y(t)$. The input to "Integrate $x''(t)$" is $\ddot{x}$ with output $\dot{x}$, which is input to "Integrate $x'(t)$", whose output will be $x(t)$. Now $x(t)$ goes to "XY Graph" , also to "Divide" , and "f(u)". "f(u)" computes $((x(t))^2+((y(t))^2)^{1.5}$ for input to "Divide" which computes $x(t)$ divided by the output of "f(u)". Then "GainK(1)" multiplies by the constant $-k$ for input to "Integrate $x''(t)$". The input to "Integrate $y''(t)$" is $\ddot{y}$ with output $\dot{y}$ for "Integrate $y'(t)$" whose output is $y(t)$. Now y goes to "XY Graph" , and also to "f(u)", and "Divide". "Divide" gets y divided by the output of "f(u)". "GainK(2)" multiplies by constant $-k$. "XY Graph" graphs x versus y to show an elliptic orbit. However, "To File" gathers all the data so we can graph all the orbits on one coordinate system. More details on this are in Chapter 28. The initial conditions are set inside the "Integrate" boxes. There are also other items to set/choose, like simulation time, type of numerical integrator and step size, which we will not discuss here (see Chapter 28).

Let $\overline{z}(t)[0] = [z_1(t,0), z_2(t,0)]$. In Figure 11.3 the outer boundary is our approximation to $z_2(t,0)$ and the inner boundary is an approximation to

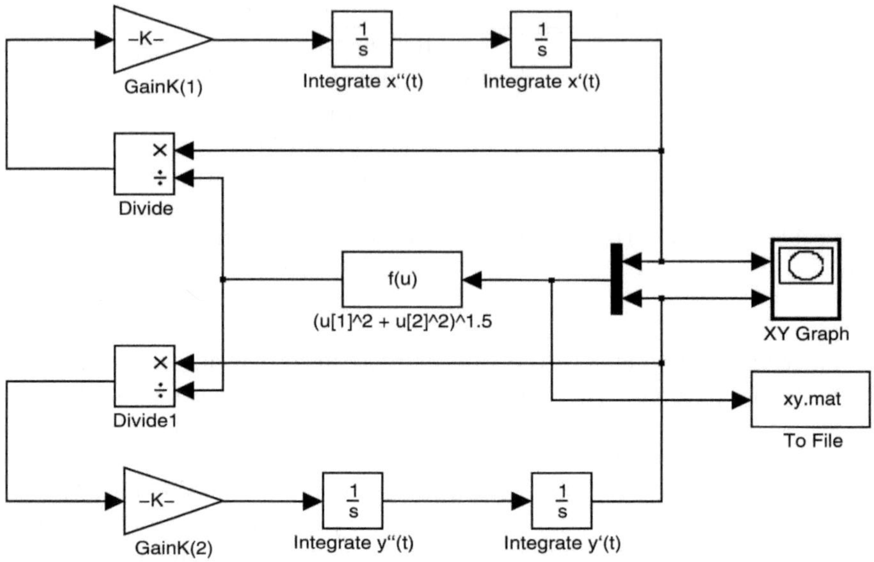

Figure 11.2: Simulink Diagram for the Earth's Orbit

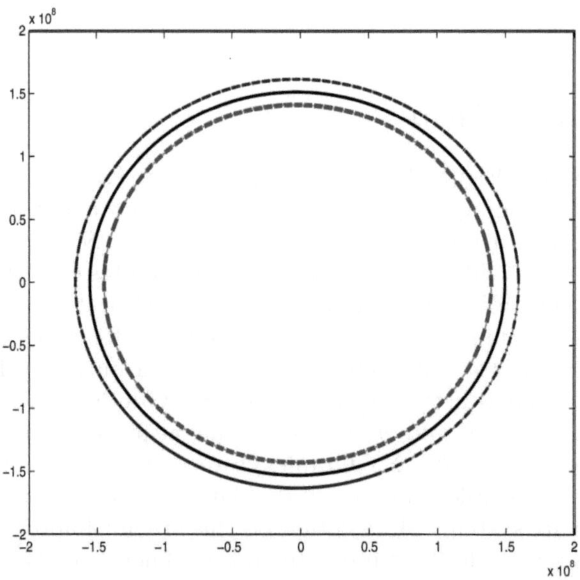

Figure 11.3: Fuzzy Trajectory for $\alpha = 0$ Cut of the Earth's Orbit

$z_1(t,0)$. The horizontal axis is x and the vertical axis is y. The fuzzy orbit looks quite stable. The uncertainty band, the difference between the two curves does not look excessive. Using our fuzzy mass of the sun and a fuzzy perihelion does not greatly change the orbit of the earth around the sun. In fact we get essentially three curves for the three values for a.

If all the uncertain parameters are estimated from data (Chapter 3), instead of employing expert opinion, then the base of the fuzzy estimator is a 99% confidence interval. Using all of these confidence intervals we get $\overline{z}(t)[0]$, so this band of uncertainty will be like 99% confidence intervals for $z(t)$.

11.4 References

1. R.L.Finney and G.B.Thomas Jr.: Calculus, Second Edition, Addison-Wesley, Reading, Mass., 1994.

2. Solar System, Microsoft Encarta Encyclopedia, Microsoft Corporation, 1995.

Chapter 12

Human Cannon Ball

12.1 Introduction

This chapter is about the human cannon ball show at the circus. A person is shot out of a cannon, flies through the air, and lands on some net. The trajectory is shown in Figure 12.1. This example is adapted from a simulation problem discussed in [1]. The "Max" on the horizontal axis is the maximum range of the human cannon ball before he/she ends up in the stands (where the customers sit). Assume that the horizontal Max is 300 feet. The "Max" on the vertical scale is the height of the roof where this circus will perform and the human cannon ball's trajectory must be below this Max line. Assume that the vertical Max is 100 feet.

From Figure 12.1 there are two forces acting on the human cannon ball: (1) (vector) $\overrightarrow{F}$ = air resistance; and (2) $w = mg$ the person's weight. Let $|\overrightarrow{F}| = F$. Using force $= ma$ the initial system of differential equations is

$$m\ddot{x} = -F\cos(\theta), \tag{12.1}$$

$$m\ddot{y} = -mg - F\sin(\theta), \tag{12.2}$$

where $\ddot{x}$ and $\ddot{y}$ is the second derivative on time for x and y, respectively. Also, in Figure 12.1 $\overrightarrow{V}$ is the tangent vector to the curve representing velocity, the direction of the vector $\overrightarrow{F}$ is $-\overrightarrow{V}$, m is the mass of the human cannon ball,w is the person's weight in pounds and $g = 32\ ft^2/sec$. Now

$$|\overrightarrow{V}| = V = \sqrt{(\dot{x})^2 + (\dot{y})^2}. \tag{12.3}$$

Assume that the initial velocity vector is $\overrightarrow{V}_0$ and $|\overrightarrow{V}_0| = V_0$ in feet per second and the initial angle of the cannon, with the horizontal, is θ_0. Let the rest of the initial conditions be: (1) $x(0) = 0$; (2) $y(0) = 0$; (3) $\dot{x}(0) = V_0\cos(\theta_0)$; and (4) $\dot{y}(0) = V_0\sin(\theta_0)$. From Figure 12.1 we see that the tangent line to the graph is assumed to make an angle θ with the horizontal.

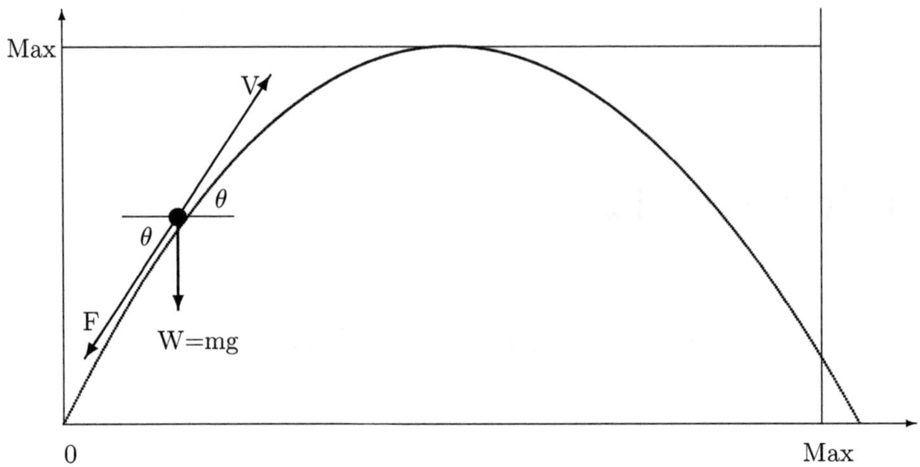

Figure 12.1: Human Cannon Ball Example

We may simplify the system of differential equations to

$$\ddot{x} = -(F/m)(\dot{x}/V), \tag{12.4}$$

$$\ddot{y} = -g - (F/m)(\dot{y}/V). \tag{12.5}$$

Now we need to make our assumptions about air resistance. The model we will use is

$$F = cV^{\beta}, \tag{12.6}$$

for constant $c > 0$ and constant $\beta \geq 1$. Substituting for F in equations (12.4) and (12.5) we obtain the final system of nonlinear differential equations to solve

$$\ddot{x} = -\frac{c}{m}V^{\beta-1}\dot{x}, \tag{12.7}$$

$$\ddot{y} = -g - \frac{c}{m}V^{\beta-1}\dot{y}. \tag{12.8}$$

Let $x(t)$ solve equation (12.7) and assume $y(t)$ solves equation (12.8). Let $z(t) = (x(t), y(t))$ represent the trajectory of the human cannon ball.

12.2 Parameters

The parameters in the model are c, β, V_0, w and θ_0. Assume that the weight of the human cannon ball is known exactly (165 pounds) and the initial angle θ_0 can be measured precisely (35 degrees) but the other parameter values are fuzzy. Through expert advice (Chapter 3) we obtain triangular fuzzy number estimators for c, β and V_0. All the parameter values, fuzzy and crisp, are given in Table 12.1. Using fuzzy parameters we get fuzzy differential equations and

Parameter	Crisp/Fuzzy Value
w=weight	165 pounds
θ_0	35 degrees
c	$\overline{c} = (1.5/2/2.5)$
β	$\overline{\beta} = (1.0/1.2/1.4)$
V_0	$\overline{V}_0 = (125/150/175)$

Table 12.1: Crisp/Fuzzy Parameters in the Human Cannon Ball Model

the solution $\overline{z}(t)$ is also fuzzy. The system now becomes a continuous fuzzy system whose trajectories are fuzzy so that any slice through a trajectory at some time t_z is a fuzzy number. We wish to estimate the band of uncertainty which is $\overline{z}(t)[0]$, or the $\alpha = 0$ cuts.

Next we need to choose the values of the fuzzy parameters in their $\alpha = 0$ cut to approximate the outer boundary of the $\alpha = 0$ cut of the fuzzy trajectory for the human cannon ball. As discussed in Chapter 6 if $\overline{\omega} = (\omega_1/\omega_2/\omega_3)$ is a fuzzy parameter we use $\omega = \omega_1$, ω_2 and ω_3. We do this for $\overline{c}$, $\overline{\beta}$ and $\overline{V}_0$ giving $3^3 = 27$ curves for $z(t)$ for a minimal approximation to $\overline{z}(t)[0]$. This is shown in Figure 12.3. The systems diagram for Simulink is in Figure 12.2.

In Figure 12.3 the reader can determine, from consulting Section 1.5.4, $z(t)$ when we used for the fuzzy parameters: (1) all left end points of their $\alpha = 0$ cut; (2) all right end points of their $\alpha = 0$ cut; and (3) their $\alpha = 1$ values.

12.3 First Simulation

Let us first discuss the Simulink diagram in Figure 12.2. The top loop is for $x(t)$ and the lower loop is for $y(t)$. The input to "Integrate x''" is $\ddot{x}$ with output $\dot{x}$. $\dot{x}$ goes to "Integrate x'" whose output is x and it also travels to "f(u)" and "Product of Elements1". Once x is calculated it travels to "XY Graph". In the lower loop $\ddot{y}$ enters "Integrate y''" with output $\dot{y}$ which is sent to "Integrate y'" and also to "f(u)" and "Product of Elements2". The output of "Integrate y'" is y which goes to "XY Graph". Now the first "f(u)" computes V which, together with β, go into the second "f(u)". This second "f(u)" computes $-(V^{\beta-1})/m$ which gets multiplied by c in "GainC" and sent on to both "Product of Elements". In the top loop "Product of Elements1" outputs $\ddot{x}$ and in the lower loop "Product of Elements2" outputs $\ddot{y}$ after subtracting g. The "XY Graph" only does one graph at a time so the "To XY File" collects all the data so we can present all the graphs in one coordinate system. More on this in Chapter 28. The initial values of x and y are set inside the "Integrators". There are also other items to set/choose, like

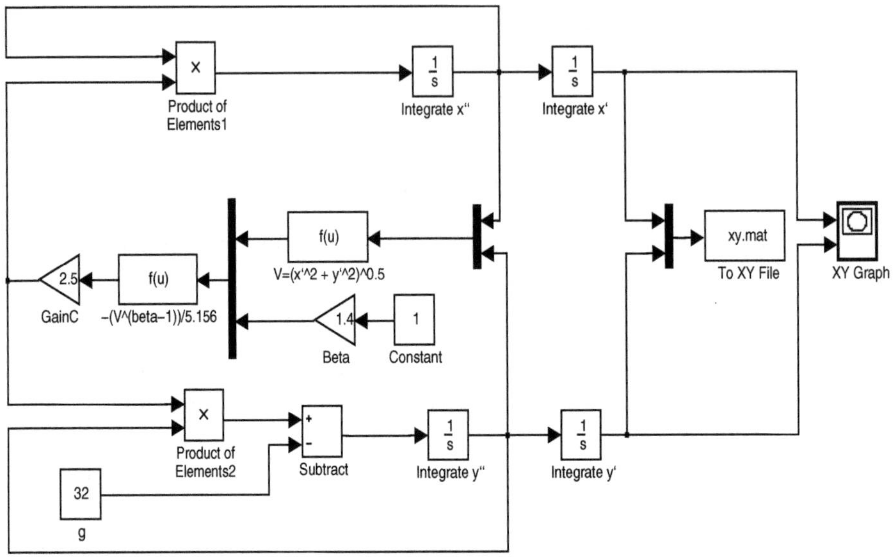

Figure 12.2: Simulink Diagram for the Human Cannon Ball Example

simulation time and type of numerical integrator, which we will not discuss here (see Chapter 28).

Let $\overline{z}(t)[0] = [z_1(t,0), z_2(t,0)]$. In Figure 12.3 connect the upper boundary with a smooth curve which is our approximation to $z_2(t,0)$ and connect the lower boundary with a smooth curve producing an approximation to $z_1(t,0)$. The horizontal axis is distance from the cannon and the vertical axis is the distance above the floor. Since the vertical Max is 100 feet there is a possibility that the human cannon ball will hit the roof. But, the human cannon ball can definitely land in the stands since the horizontal Max is 300 feet. We need to redesign this act.

12.4 Second Simulation

All we change is the fuzzy initial velocity. Now assume that $\overline{V}_0 = (80/100/120)$ with no other changes. There are different settings on the cannon so we can easily alter the initial velocity. The results are in Figure 12.4. The human cannon ball now misses the roof and does not land in the audience.

If all the uncertain parameters are estimated from data (Chapter 3), and not from expert advice, then the base of each fuzzy estimator is a 99% confidence interval. Using all of these confidence intervals we get $\overline{z}(t)[0]$, so these bands of uncertainty will be like 99% confidence intervals for $z(t)$.

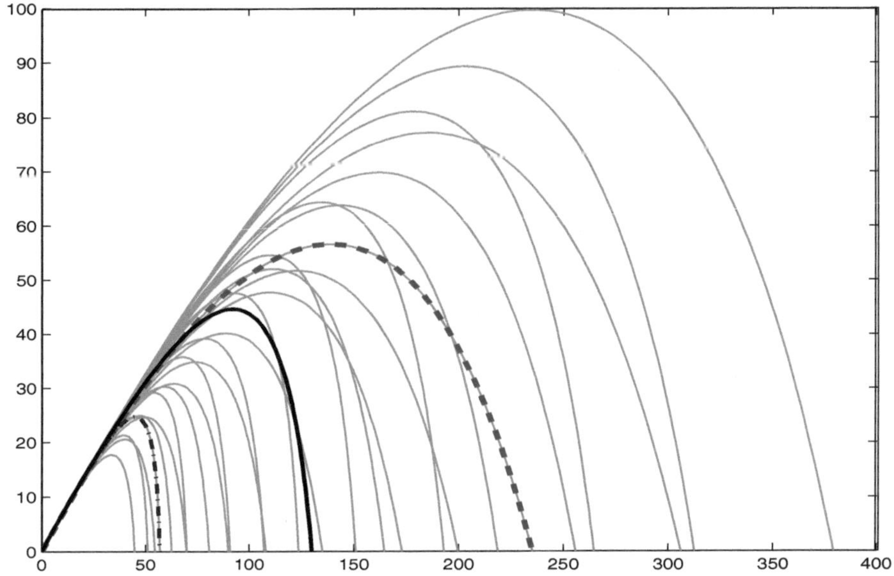

Figure 12.3: Fuzzy Trajectory for the Human Cannon Ball Example: First Simulation

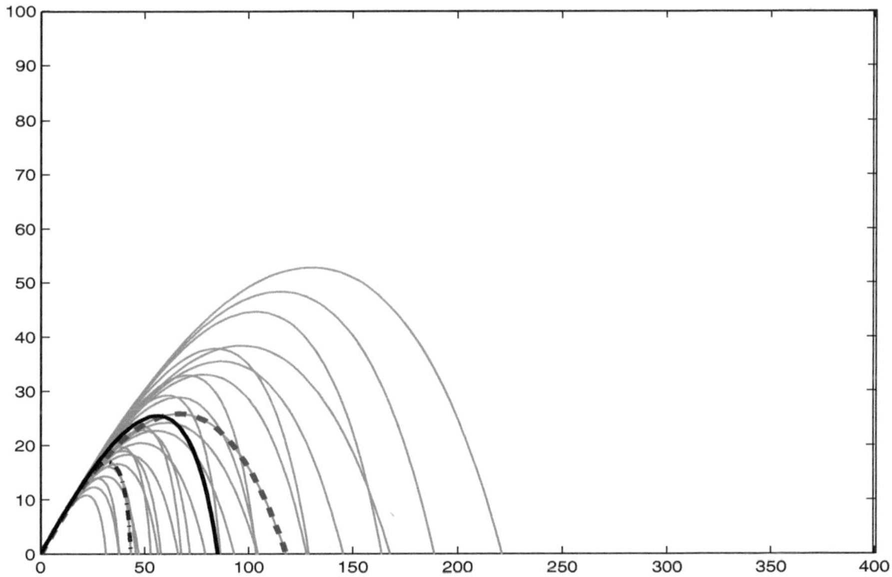

Figure 12.4: Fuzzy Trajectory for the Human Ball Example: Second Simulation

12.5 Reference

1. G.A.Korn and J.V.Wait: Digital Continuous-System Simulation, Prentice-Hall, Englewood Cliffs, NJ, 1978.

Chapter 13

Electrical Circuits

13.1 Introduction

The simple electrical network we shall study in this chapter is shown in Figure 13.1. In this figure C_i stands for a capacitor, R_i denotes a resistor, L_i is an inductance and E_i represents an electromotive force, or some type of alternating current input to the system. We will travel around both loops in the circuit in a counter clockwise direction. This application was adopted from an example in [1]. In order to get the system of differential equations for this diagram we need to review (learn) some facts about simple electrical networks.

Let Q_i be the charge (coulombs) on capacitor C_i, $i = 1, 2$. The currents I_i (amperes) shown in Figure 13.1 are related to the charges as $dQ_i/dt = I_i$ for $i = 1, 2$. Let us consider the loop on the left side of Figure 13.1. By Kirchhoff's Second Law, the impressed voltage E_1 must equal the sum of the voltage drops in the loop. The voltage drop across the inductance L_1 (henrys) is $L_1(dI_1/dt)$. The voltage drop across the resistance R_1(ohms) is I_1R_1 and the voltage drop across the capacitance C_1(farads) is Q_1/C_1. Notice we have $I_1 = I_2 + I_3$. Then

$$E_1 = L_1\ddot{Q}_1 + R_1\dot{Q}_1 + (1/C_1)Q_1 + R_3I_3, \tag{13.1}$$

where $\ddot{Q}$ ($\dot{Q}$) is the second (first) derivative of Q with respect to time and $\dot{Q} = I$. Using $I_3 = I_1 - I_2$ we have our first equation

$$E_1 = L_1\ddot{Q}_1 + R_1\dot{Q}_1 + (1/C_1)Q_1 + R_3(\dot{Q}_1 - \dot{Q}_2). \tag{13.2}$$

For the loop on the right side of Figure 13.1 we have the second equation

$$E_2 = L_2\ddot{Q}_2 + R_2\dot{Q}_2 + (1/C_2)Q_2 + R_3(\dot{Q}_2 - \dot{Q}_1). \tag{13.3}$$

The initial conditions will be $Q_i(0) = 0$, $\dot{Q}_i(0) = 0$, $i = 1, 2$.

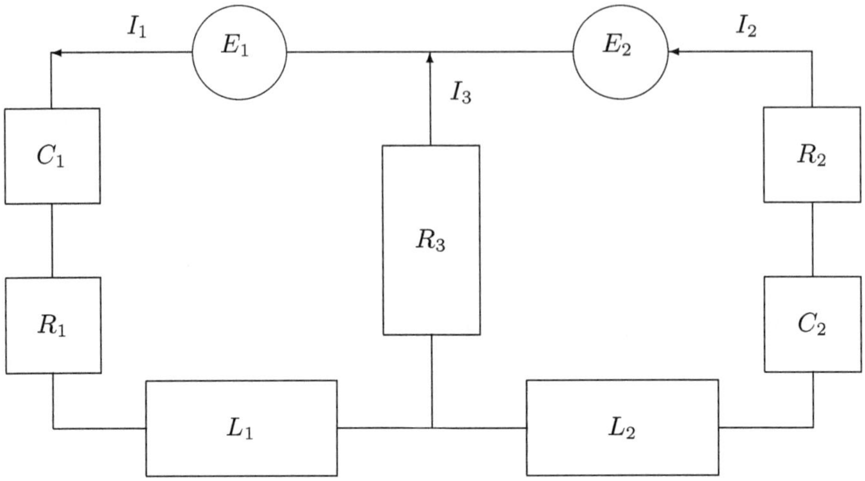

Figure 13.1: Electrical Network

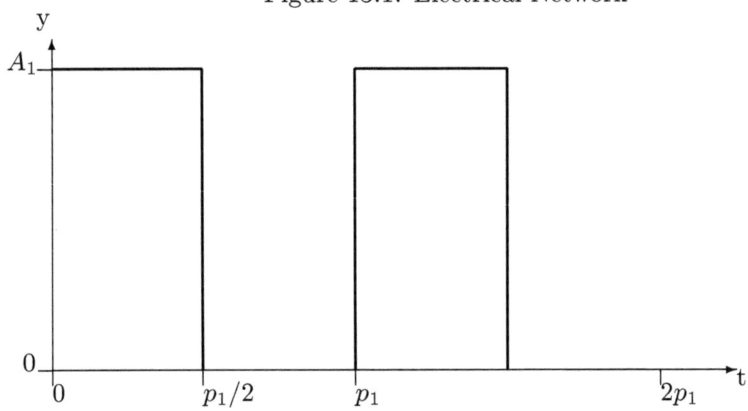

Figure 13.2: Square Wave Input E_1

We have a linear system of differential equations to solve for the currents I_i, $i = 1, 2, 3$. We could find the solutions which will be fairly complicated because of what we shall use for E_1 and E_2. However, even using simple solutions it can be difficult to find the boundary of the band of uncertainty, or the boundary of $\overline{I}_i(t)[0])$, $i = 1, 2, 3$, as shown in Section 6.2. So we omit expressions for the exact solutions and will use simulation instead.

13.2 Parameters

All the parameters in these equations are assumed known and crisp, and are given in Table 13.1. What is fuzzy in this problem are the energy sources E_1 and E_2 (see Figures 13.2 and 13.3). E_1 will be a square wave with amplitude

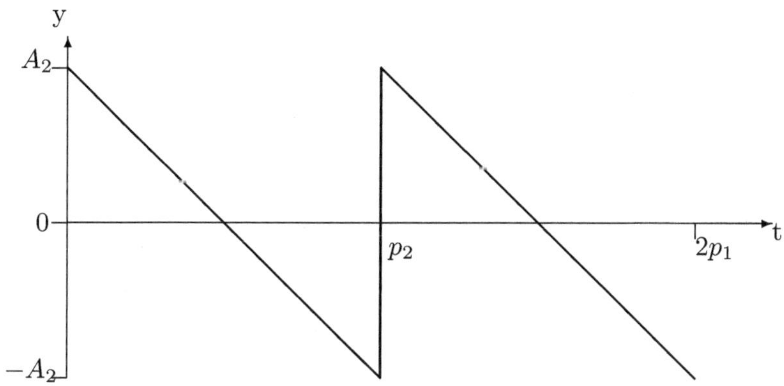

Figure 13.3: Saw Tooth Input E_2

Parameter	Crisp/Fuzzy Value
L_1	0.25 henrys
L_2	0.5 henrys
C_1	0.1 farads
C_2	0.3333 farads
R_1	10 ohms
R_2	5 ohms
R_3	20 ohms
A_1	$\overline{A}_1 = (2.25/2.5/2.75)$
A_2	$\overline{A}_2 = (4.5/5/5.5)$
p_1	$\overline{p}_1 = (2.7/3/3.3)$
p_2	$\overline{p}_2 = (1.5/2/2.5)$

Table 13.1: Crisp/Fuzzy Parameter Values in the Electrical Network

A_1 and period p_1 and E_2 is a sawtooth wave with amplitude A_2 and period p_2. We will use fuzzy numbers for the amplitudes and periods and they are given in Table 13.1. So the energy sources will be fuzzy.

Using fuzzy parameters for amplitudes and periods we get fuzzy differential equations and the solutions $\overline{I}_i(t)$ are also fuzzy. The system now becomes a continuous fuzzy system whose trajectories are fuzzy so that any slice through a trajectory at some time t_z is a fuzzy number. We wish to estimate the bands of uncertainty which are $\overline{I}_i(t)[0]$, or the $\alpha = 0$ cuts.

Next we need to choose the values of the fuzzy parameters in their $\alpha = 0$ cut to approximate the outer boundary of the $\alpha = 0$ cut of the fuzzy trajectory for the currents. As discussed in Chapter 6 if $\overline{\omega} = (\omega_1/\omega_2/\omega_3)$ is a fuzzy parameter we use $\omega = \omega_1$, ω_2 and ω_3. We do this for $\overline{A}_i$ and $\overline{p}_i$, $i = 1, 2$, giving $3^4 = 81$ curves for $I_i(t)$, for a minimal approximation to $\overline{I}_i(t)[0]$, $i = 1, 2, 3$. These are shown in Figures 13.5 - 13.7. The systems diagram for Simulink is in Figure 13.4.

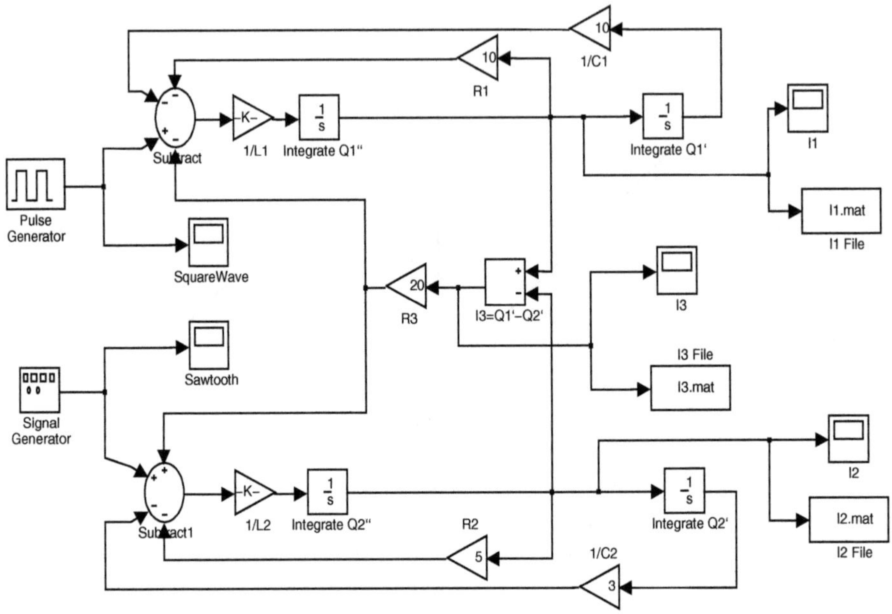

Figure 13.4: Simulink Diagram for the Electrical Network

In Figure 13.5 (13.6,13.7) the reader can determine, from consulting Section 1.5.4, $I_1(t)$ ($I_2(t)$,$I_3(t)$) when we used for the fuzzy parameters: (1) all left end points of their $\alpha = 0$ cut; (2) all right end points of their $\alpha = 0$ cut; and (3) their $\alpha = 1$ values.

13.3 Simulation

We will first discuss the Simulink diagram. The upper loop is for I_1, the middle gets I_3 and the bottom loop computes I_2. The input to "Integrate $Q1''$" is $\ddot{Q}_1$ with output $\dot{Q}_1$ which is input to "Integrate $Q1'$" whose output is Q_1. $\dot{Q}_1$ get multiplied by R_1 before input to "Subtract". Also Q_1 gets multiplied by $\frac{1}{C_1}$ before input to "Subtract". $\dot{Q}_1$ is also sent to "$I_3 = Q_1' - Q_2'$" which calculates $\dot{Q}_1 - \dot{Q}_2$ and then is multiplied by R_3 before input to "Subtract" and to "Subtract1". The "Pulse Generator" makes the square wave for E_1 which goes to "Subtract". "Subtract" gets multiplied by $1/L_1$, and we now have $\ddot{Q}_1$ obtained from equation (13.2), after we first solve that equation for $\ddot{Q}_1$. Now $\dot{Q}_1 = I_1$ travels to a Scope called "I_1" for the graphing of this current. However, a Scope only does one graph at a time and "I_1 File" gathers all the data so that we can graph all the 81 curves on one coordinate -5.3 amps to a maximum of around 6 amps.

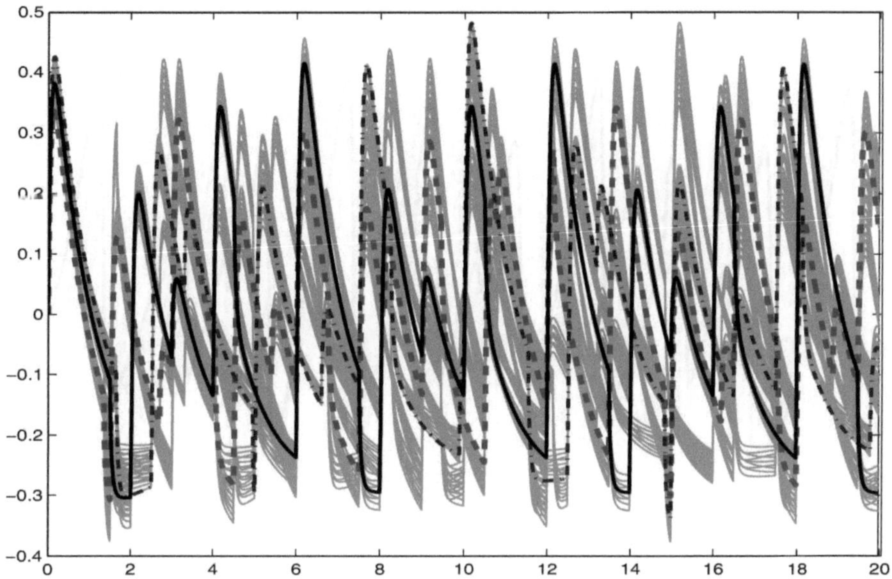

Figure 13.5: Fuzzy Trajectory for Fuzzy Current $\overline{I}_1$

system. More details on how to accomplish this is in Chapter 28. A similar description holds for the bottom loop. The "Signal Generator" generates the sawtooth wave for E_2 which travels to "Subtract1". "Subtract1" gets multiplied by $1/L_2$ and we have $\ddot{Q}_2$ from equation (13.3), after we first solve that equation for $\ddot{Q}_2$. We get I_3 from $\dot{Q}_1 - \dot{Q}_2 = I_1 - I_2 = I_3$ which is sent to Scope called "I_3" and "I_3 File" collects all the graphs to put into one figure. We also see that $\dot{Q}_2 = I_2$ goes to Scope "I_2" and "I_2 File". There are also other items to set/choose, like using "time based" or "sample based" time in the E_i, using "fixed time" or "variable time" steps in the "Integrators", which we will not discuss here (see Chapter 28).

Let $\overline{I}_1(t)[0] = [I_{11}(t,0), I_{12}(t,0)]$. In Figure 13.5 connect the upper boundary with a smooth curve which is our approximation to $I_{12}(t,0)$ and connect the lower boundary with a smooth curve producing an approximation to $I_{12}(t,0)$. The horizontal axis is time and the vertical axis is the current in amps. This is quite a complicated picture. We see that I_1 oscillates back and forth between a maximum of about 4.9 amps and a minimum of approximately -3.9 amps.

Let $\overline{I}_2(t)[0] = [I_{21}(t,0), I_{22}(t,0)]$. In Figure 13.6 connect the upper boundary with a smooth curve which is our approximation to $I_{21}(t,0)$ and connect the lower boundary with a smooth curve producing an approximation to $I_{22}(t,0)$. The horizontal axis is time and the vertical axis is the current in amps. Another complicated figure. I_2 can oscillate from a minimum of about

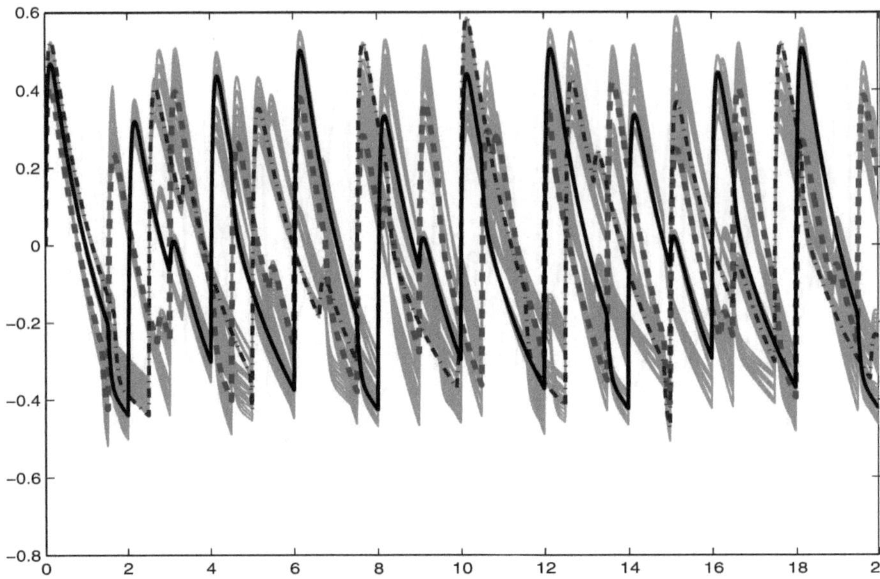

Figure 13.6: Fuzzy Trajectory for Fuzzy Current $\overline{I}_2$

Figure 13.7: Fuzzy Trajectory for Fuzzy Current $\overline{I}_3$

Let $\overline{I}_3(t)[0] = [I_{31}(t,0), I_{32}(t,0)]$. In Figure 13.7 connect the upper boundary with a smooth curve which is our approximation to $I_{31}(t,0)$ and connect the lower boundary with a smooth curve producing an approximation to $I_{32}(t,0)$. The horizontal axis is time and the vertical axis is the current in amps. I_3 may vary between approximately -0.175 amps to 0.2 amps.

Fuzzy energy inputs produces complicated and very fuzzy currents.

If all the uncertain parameters are estimated from data (Chapter 3), and not employing expert opinion, then the base of the fuzzy estimator is a 99% confidence interval. Using all of these confidence intervals we get $\overline{I}_i(t)[0]$, $i = 1,2,3$, so these bands of uncertainty will be like 99% confidence intervals for the $I_i(t)$, $i = 1,2,3$.

13.4 Reference

1. W.Kaplan: Ordinary Differential Equations, Addison-Wesley, Reading, Mass., 1962.

Chapter 14

Hawks, Doves and Law-Abiders

14.1 Introduction

This example is adopted from [1] where the author attributes the example to [2] where that author tried "to develop a formal model of the Hobbesian state of nature from the perspective of bioeconomics". In [2] the author defines Hobbes' state of nature as a society whose members are competing with each other to obtain resources. All resources belong to someone so conflicts arise between the resource owners and those who want the resource. The author divides the members of society into three groups according to their strategies: the hawk; the dove; and the law-abiding citizen. We have the definitions:

1. the dove never tries to get others' possessions. He waits until they give up their resource and he himself abandons his resource if attacked. If two doves compete for the same resource one gets it with equal probability.

2. the hawk always tries to get others' resources by means of aggression and gives up his resources only if seriously injured.

3. the law-abider never tries to get others' resources. He waits until they give up their resources. He defends his resources and losses them only if defeated.

Let $x(t)$ ($y(t)$, $z(t)$) denote the proportion of doves (hawks, law-abiders) in the society at time t. So, $x(t)$, $y(t)$ and $z(t)$ all belong to $[0, 1]$ with sum one. After a lot of discussion and calculations [1] finally ends up with the following system of nonlinear differential equations to solve

$$\dot{x} = 4x^2 - 4x + 9xy - 13x^2y, \tag{14.1}$$

Initial Condition	First Simulation	Second Simulation	Third Simulation
$\overline{x}(0) = Doves$	(0.01/0.05/0.09)	(0.494/0.499/0.504)	(0.97/0.98/0.99)
$\overline{y}(0) = Hawks$	(0.91/0.95/0.99)	(0.494/0.499/0.504)	(0.005/0.015/0.025)
$z(0) = Abiders$	0	0.002	0.005

Table 14.1: Fuzzy/Crisp Initial Conditions

$$\dot{y} = 2.5y^2 - 2.5y + 10.5xy - 13xy^2, \tag{14.2}$$

$$z = 1 - x - y. \tag{14.3}$$

The initial conditions are $x(0) = x_0$, $y(0) = y_0$ and $z(0) = z_0 = 1 - x_0 - y_0$. The notation $\dot{x}$ is the time derivative of $x(t)$, etc. We will not go through all the details of deriving these differential equations but instead refer the reader to [1] for more information.

14.2 Parameters

We will accept all the crisp parameter values in the system of differential equations (none of them will be fuzzy), but the initial conditions are uncertain and will be fuzzy. Let us use for our first simulation triangular fuzzy numbers $\overline{y}(0) = (0.91/0.95/0.99)$, $\overline{x}(0) = 1 - \overline{y}(0) = (0.01/0.05/0.09)$ and $\overline{z}(0) = 0$ (crisp zero). Table 14.1 has the fuzzy initial conditions for the various simulations. In all cases $z(0)$ will be crisp and not fuzzy. Using fuzzy initial conditions we get fuzzy differential equations and the solutions $\overline{x}(t)$, $\overline{y}(t)$ and $\overline{z}(t)$ will also be fuzzy. The system now becomes a continuous fuzzy system whose trajectories are fuzzy so that any slice through a trajectory at some time t_z is a fuzzy number. We wish to estimate the band of uncertainty which is $\overline{x}(t)[0]$, $\overline{y}(t)[0]$ and $\overline{z}(t)[0]$, or the $\alpha = 0$ cuts.

Next we need to choose the values of the fuzzy parameters in their $\alpha = 0$ cut to approximate the outer boundary of the $\alpha = 0$ cut of the fuzzy trajectory for the doves, hawks and the law-abiders. As discussed in Chapter 6 if $\overline{\omega} = (\omega_1/\omega_2/\omega_3)$ is a fuzzy parameter we usually use $\omega = \omega_1$, ω_2 and ω_3. We would do this for $\overline{x}_0$, $\overline{y}_0$ and z_0 giving $3^2 = 9$ curves for $x(t)$, $y(t)$ and $z(t)$ for a minimal approximation to $\overline{x}(t)[0]$, $\overline{y}(t)[0]$ and $\overline{z}(t)[0]$. However, we can not do it this way in this chapter because we will honor the hard (crisp) constraint that the sum of x, y and z always equals one. Consider the first simulation. We will choose $x_0 = 0.01, 0.02, 0.03, 0.04, 0.05, 0.06, 0.07, 0.08, 0.09$ with corresponding $y_0 = 0.99, 0.98, 0.97, 0.96, 0.95, 0.94, 0.93, 0.92, 0.91$ and $z_0 = 0$. In the first simulation $x_0 + y_0 = 1$, $z_0 = 0$, with $x_0 \in [0.01, 0.09]$ and $y_0 \in [0.91, 0.99]$. In the second simulation we choose $x_0 = 0.494, 0.495, 0.496, 0.497, 0.498, 0.499, 0.500, 0.501, 0.502,$
$0.503, 0.504$ and the corresponding $y_0 = 0.998 - x_0$ so that $x_0 + y_0 + z_0 = 1$. The the last simulation let $x_0 = 0.97, 0.975, 0.98, 0.985, 0.99$ with $y_0 =$

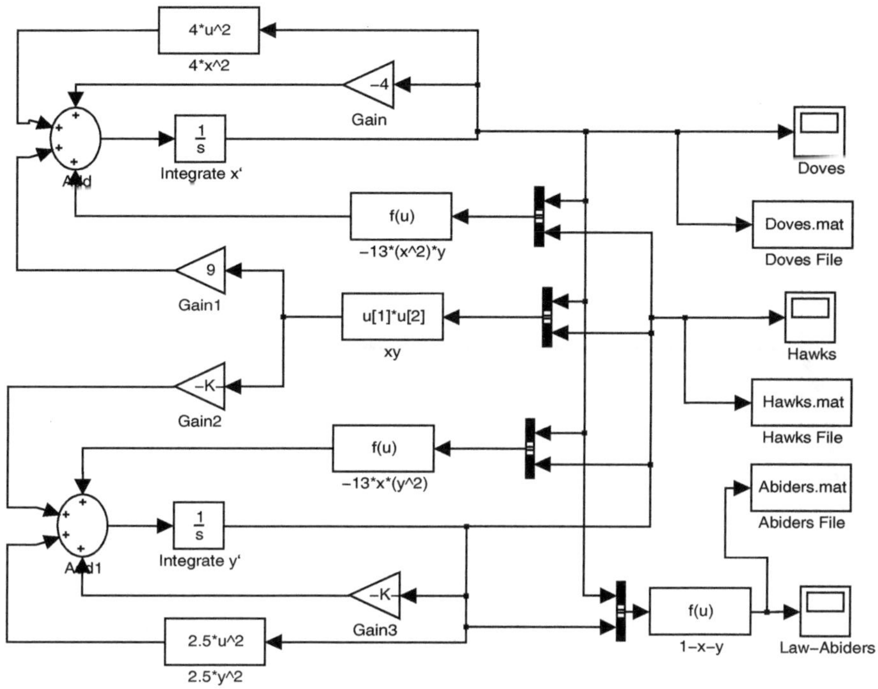

Figure 14.1: Simulink Diagram for the Hawks, Doves and Law-Abiders Example

$0.995 - x_0$ and we get the sum always one again. So the figures for the three cases will have 9,11 and 5 graphs. More details on putting multiple graphs on one coordinate system is in Chapter 28. The fuzzy graphs in the first simulation are in Figures 14.2 - 14.3. No graph for $z(t)$ is needed in the first simulation because it is always zero. The systems diagram for Simulink is in Figure 14.1. The other simulations are in Figures 14.4 - 14.9.

In Figures 14.2-14.9 the reader can determine, from consulting Section 1.5.4, $x(t)$ ($y(t)$,$z(t)$) when we used for the fuzzy parameters: (1) all left end points of their $\alpha = 0$ cut; (2) all right end points of their $\alpha = 0$ cut; and (3) their $\alpha = 1$ values.

14.3 First Simulation

We first discuss Simulink in Figure 14.1. You can see that the top loop is for the doves ($x(t)$), the middle computes the hawks ($y(t)$), and the bottom determines the law-abiders ($z(t)$). "Integrate x' " gets $x(t)$ from $\dot{x}$. "Add" computes the right side of equation (14.1). So $x(t)$ goes to "Gain" to be

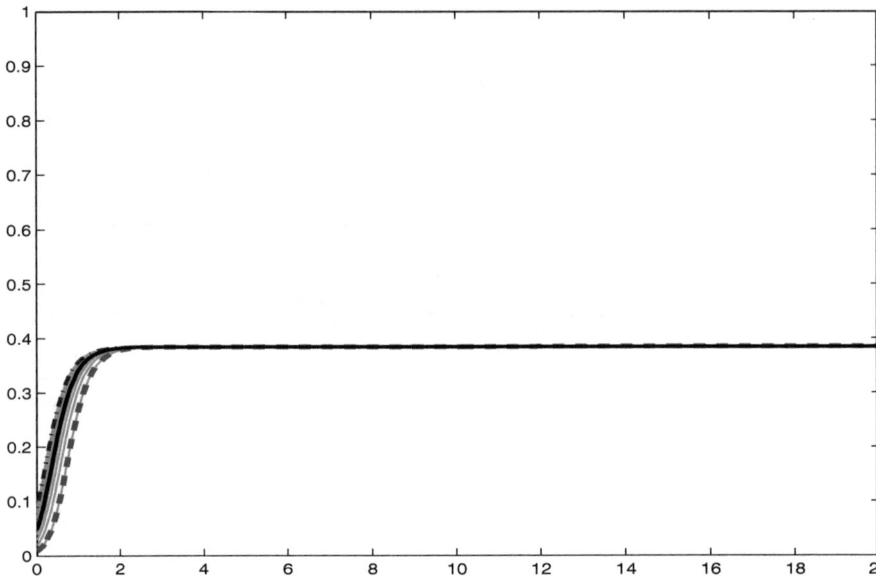

Figure 14.2: Fuzzy Trajectory for Doves $\overline{x}(t)[0]$ in the 1st Simulation

multiplied by -4, and also travels to the box that finds $4x^2$, with both input to "Add". $x(t)$ also goes to three boxes that compute $-13x^2y$ for input to "Add", xy to be multiplied by 9 in "Gain1" and $-13xy^2$ for input to "Add1". The Scope called "doves" is to graph $x(t)$ and the "Doves File" collects all the data so that all the graphs can be presented on one coordinate system. The details for the hawks ($y(t)$) should now be clear. The box at the bottom right gets $z = 1 - x - y$ for the law-abiders to be graphed in the scope called "Law-Abiders". The "Abiders File" is for all the graphs in a single coordinate system. More details on how this is done is in Chapter 28. There are also other items to set/choose, like simulation time and type of numerical integrator, etc. which we do not discuss in detail here (see Chapter 28).

It appears, from many simulations, that: (1) If $x_0 = 0$, then $x(t) = 0$ all t and the same is true for y_0 and z_0; and (2) if $z_0 > 0$, then $x(t) \to 0$, $y(t) \to 0$, $z(t) \to 1$ as $t \to \infty$. These results may not show for all curves in the third simulation because $0 \leq t \leq 20$.

Let $\overline{x}(t)[0] = [x_1(t,0), x_2(t,0)]$ for the doves. In Figure 14.2 connect the upper boundary with a smooth curve which is our approximation to $x_2(t,0)$ and connect the lower boundary with a smooth curve producing an approximation to $x_1(t,0)$. The horizontal axis is time and the vertical axis is the percent (after multiplying by 100) of doves. We get steady state (horizontal curves) with $\lim_{t\to\infty} x_i(t,0) \approx 0.0.38$ for $i = 1, 2$. There is no uncertainty in this result.

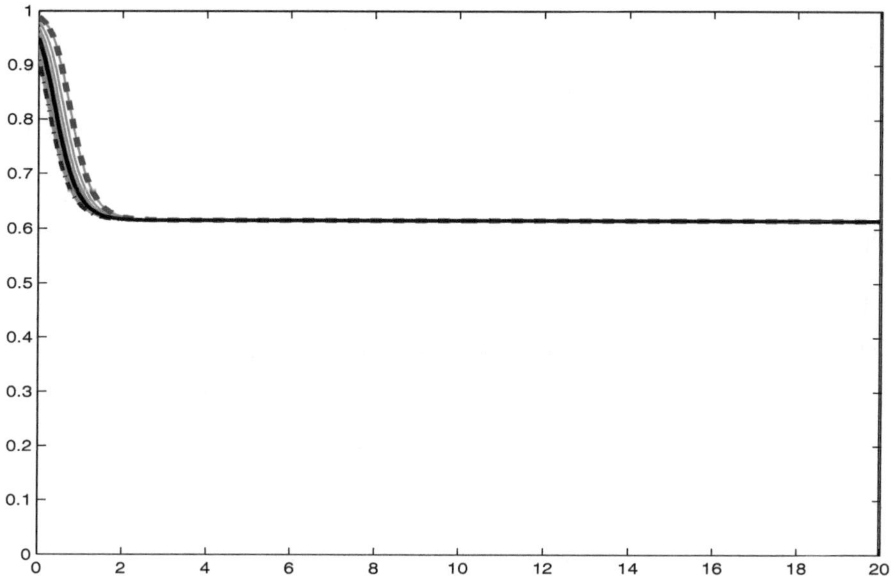

Figure 14.3: Fuzzy Trajectory for Hawks $\overline{y}(t)[0]$ in the 1st Simulation

Let $\overline{y}(t)[0] = [y_1(t,0), y_2(t,0)]$ for the hawks. In Figure 14.3 connect the upper boundary with a smooth curve which is our approximation to $y_2(t,0)$ and connect the lower boundary with a smooth curve producing an approximation to $y_1(t,0)$. The horizontal axis is time and the vertical axis is the percent (after multiplying by 100) of hawks. We get steady state (horizontal curves) with $\lim_{t\to\infty} y_i(t,0) \approx 0.62$ for $i = 1, 2$. Again, there is no uncertainty in this result.

There is no graph for the law-abiders because $z(t) = 0$ for $t \geq 0$.

14.4 Second Simulation

Now we use the fuzzy initial conditions under "Second Simulation" in Table 14.1.

Let $\overline{x}(t)[0] = [x_1(t,0), x_2(t,0)]$ for the doves. In Figure 14.4 connect the upper boundary with a smooth curve which is our approximation to $x_2(t,0)$ and connect the lower boundary with a smooth curve producing an approximation to $x_1(t,0)$. The horizontal axis is time and the vertical axis is the percent of doves (after multiplying by 100). We see that $x(t)$ begins at ≈ 0.50, then decreases rapidly to around 0.39 where it stays until time unit ≈ 160 when it immediately goes to zero. There is no uncertainty in this result. The uncertainty band is zero. This follows from very little uncertainty

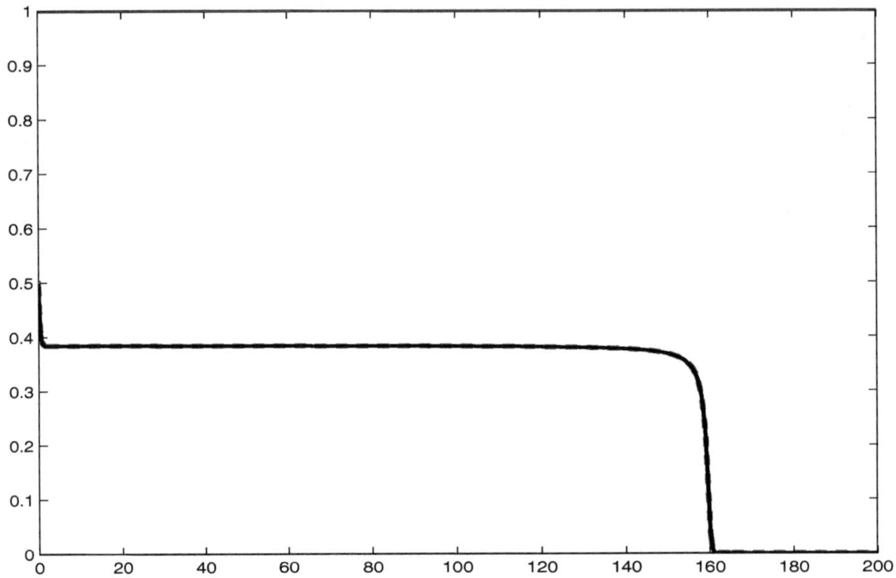

Figure 14.4: Fuzzy Trajectory for Doves $\overline{x}(t)[0]$ in the 2nd Simulation

in the initial conditions.

Let $\overline{y}(t)[0] = [y_1(t,0), y_2(t,0)]$ for the hawks. In Figure 14.5 connect the upper boundary with a smooth curve which is our approximation to $y_2(t,0)$ and connect the lower boundary with a smooth curve producing an approximation to $y_1(t,0)$. The horizontal axis is time and the vertical axis is the percent of hawks (after multiplying by 100). We see that $y(t)$ begins at ≈ 0.50, then increases rapidly to around 0.61 where it stays until time unit ≈ 160 when it immediately goes to zero. There is no uncertainty in this result. The uncertainty band is zero. This follows from very little uncertainty in the initial conditions.

Let $\overline{z}(t)[0] = [z_1(t,0), z_2(t,0)]$ for the law-abiders. In Figure 14.5 connect the upper boundary with a smooth curve which is our approximation to $z_2(t,0)$ and connect the lower boundary with a smooth curve producing an approximation to $z_1(t,0)$. The horizontal axis is time and the vertical axis is the percent of hawks (after multiplying by 100). We see that $z(t)$ begins at ≈ 0, then starts to increase slowly until around $t = 160$ when it increases rapidly to 1.00. There is no uncertainty is this result. The uncertainty band is zero. This follows from very little uncertainty in the initial conditions.

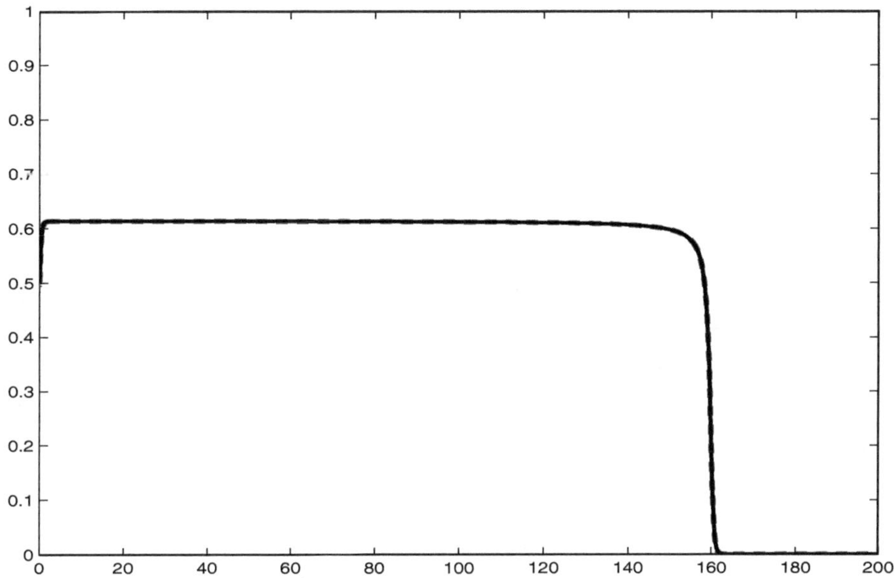

Figure 14.5: Fuzzy Trajectory for Hawks $\overline{y}(t)[0]$ in the 2nd Simulation

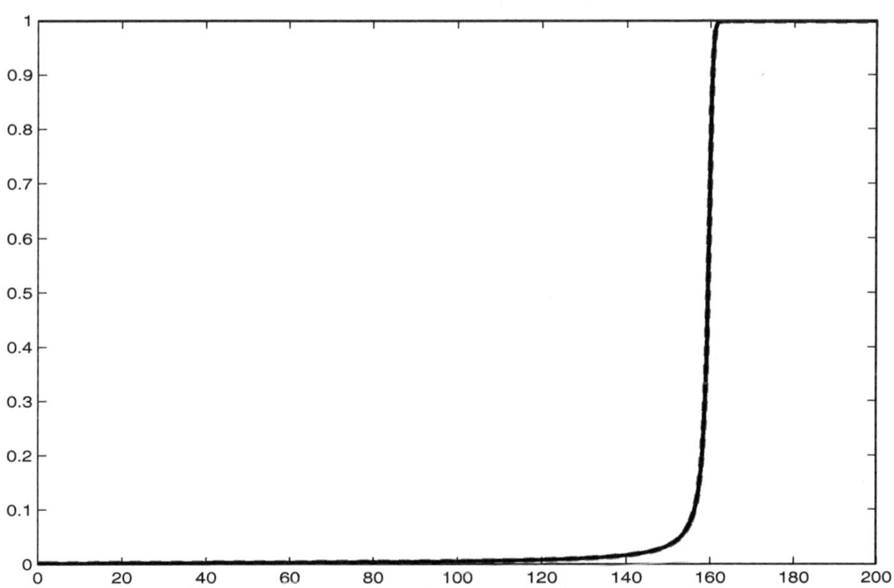

Figure 14.6: Fuzzy Trajectory for Law-Abiders $\overline{z}(t)[0]$ in the 2nd Simulation

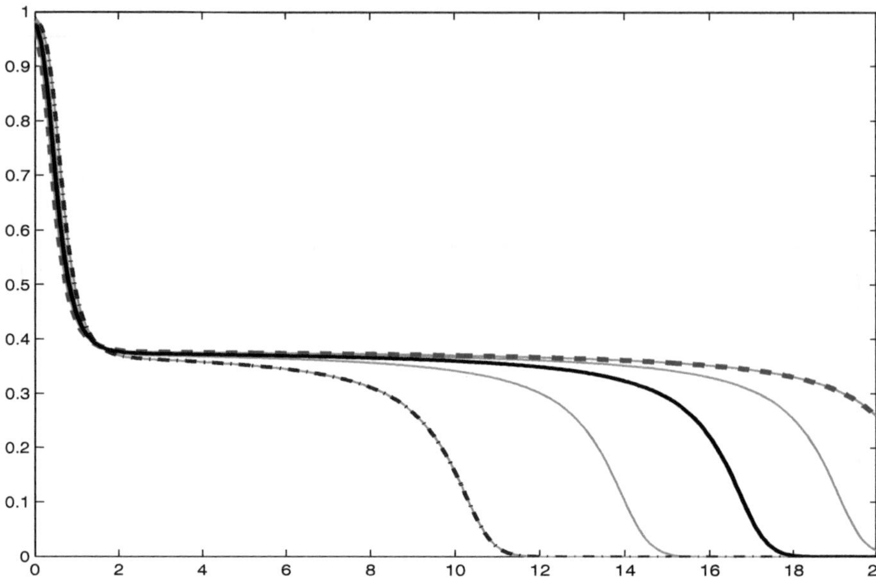

Figure 14.7: Fuzzy Trajectory for Doves $\overline{x}(t)[0]$ in the 3rd Simulation

14.5 Third Simulation

Lastly, we use the initial conditions under "Third Simulation" in Table 14.1.

Let $\overline{x}(t)[0] = [x_1(t,0), x_2(t,0)]$ for the doves. In Figure 14.7 connect the upper boundary with a smooth curve which is our approximation to $x_2(t,0)$ and connect the lower boundary with a smooth curve producing an approximation to $x_1(t,0)$. The horizontal axis is time and the vertical axis is the percent of doves (after multiplying by 100). We see that $x(t)$ begins at ≈ 0.98, then decreases to approximately 0.39 followed by decreasing to zero. There is little uncertainty in this result until the end when it goes to zero.

Let $\overline{y}(t)[0] = [y_1(t,0), y_2(t,0)]$ for the hawks. In Figure 14.8 connect the upper boundary with a smooth curve which is our approximation to $y_2(t,0)$ and connect the lower boundary with a smooth curve producing an approximation to $y_1(t,0)$. The horizontal axis is time and the vertical axis is the percent of hawks (after multiplying by 100). We see that $y(t)$ begins at ≈ 0.015, then increases rapidly to around 0.61 where it stays until it decreases to zero. There is little uncertainty in this result until it is time to go to zero.

Let $\overline{z}(t)[0] = [z_1(t,0), z_2(t,0)]$ for the law-abiders. In Figure 14.9 connect the upper boundary with a smooth curve which is our approximation to $z_2(t,0)$ and connect the lower boundary with a smooth curve producing an approximation to $z_1(t,0)$. The horizontal axis is time and the vertical axis is the percent of hawks (after multiplying by 100). We see that $z(t)$ begins at

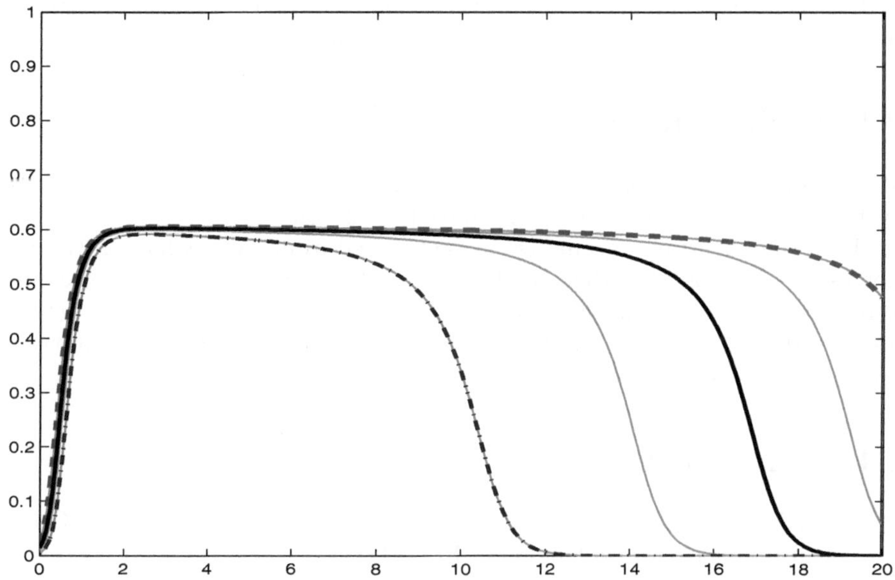

Figure 14.8: Fuzzy Trajectory for Hawks $\overline{y}(t)[0]$ in the 3rd Simulation

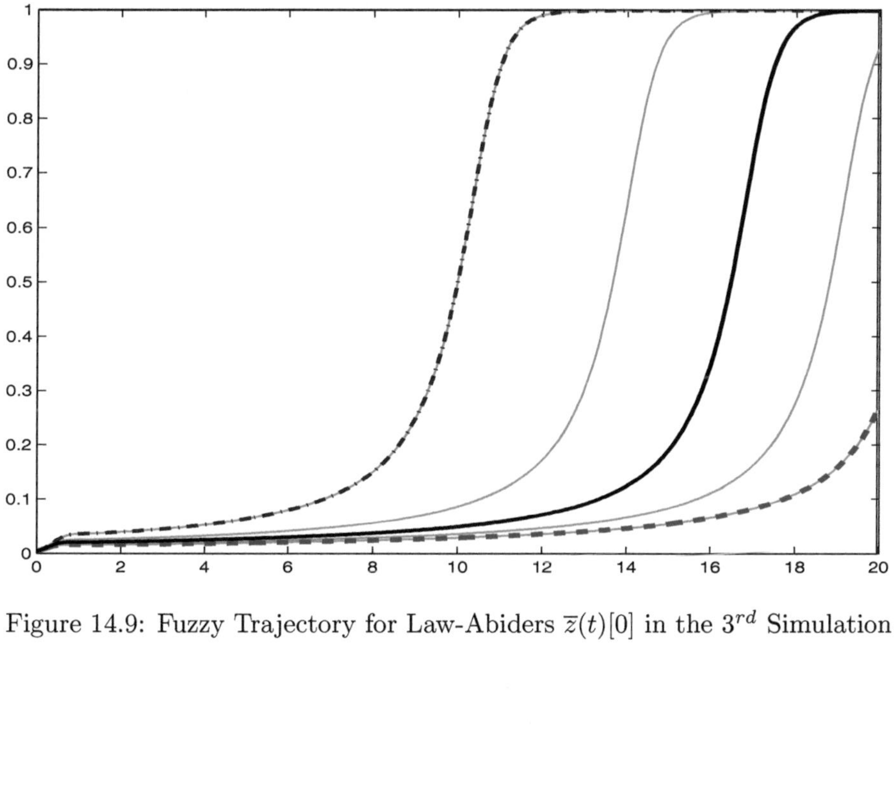

Figure 14.9: Fuzzy Trajectory for Law-Abiders $\overline{z}(t)[0]$ in the 3rd Simulation

≈ 0, then starts to increase slowly until it increases rapidly to 1.00. There is little uncertainty is this result until it goes to one.

14.6 Summary

We see quite different results in the three simulations showing that this fuzzy system is very sensitive to its fuzzy initial conditions.

If all the uncertain parameters are estimated from data (Chapter 3), and not using expert advice, then the base of the fuzzy estimator is a 99% confidence interval. Using all of these confidence intervals we get $\overline{x}(t)[0]$, $\overline{y}(t)[0]$ and $\overline{z}(t)[0]$ so these bands of uncertainty will be like 99% confidence intervals for $x(t)$, $y(t)$ and $z(t)$, respectively.

14.7 References

1. N.Gilbert and K.G.Troitzsch: Simulation for the Social Scientist, Open University Press, Philadelphia, PA, 2003.

2. J.C.Martinez Coll: A Bioeconomic Model of Hobbes' "State of Nature", Social Science Information, 25(1986)493-505.

Chapter 15

Suspension System

15.1 Introduction

This application is modeled after an example in [1]. It has to do with the design of a suspension system for a car. Figure 15.1 contains the information for this problem. The units will be force in newtons (N), time in seconds (s), distance in meters (m) and mass in kilograms (kg). M_1 is 25% the mass of the car and M_2 is the mass of one half of an axle plus a tire. k_1 is the car spring constant (N/m) and f is the shock absorber damping constant (Ns/m). Also k_2 is the tire spring constant (N/m). The car's vertical displacement is measured by x and the tire's vertical displacement is measured by y. The car moves along smoothly ($x = y = 0$) until the tire hits a "bump" in the road. This bump could be caused by crossing a train track at a rail crossing, or by a rock in the road. The vertical displacement of the tire due to the bump at time zero is measured by the constant z. We now may derive the following system of linear differential equations [1]

$$M_1\ddot{x} = -f(\dot{x} - \dot{y}) - k_1(x - y), \tag{15.1}$$

$$M_2\ddot{y} = f(\dot{x} - \dot{y}) + k_1(x - y) + k_2(z - y), \tag{15.2}$$

where $\ddot{x}$ ($\ddot{y}$) is the second derivative of x (y) with respect to time and $\dot{x}$ ($\dot{y}$) is the first derivative of x (y) with respect to time. The initial conditions are $x(0) = 0$, $y(0) = z$, $\dot{x}(0) = 0$ and $\dot{y}(0) = 0$. These are linear differential equations so we could easily find the solutions. However, even using simple solutions it can be complicated to find the boundary of the band of uncertainty, or the boundary of $\overline{x}(t)[0]$ and $\overline{y}(t)[0]$, as shown in Section 6.2. So we omit expressions for the exact solutions and will use simulation instead.

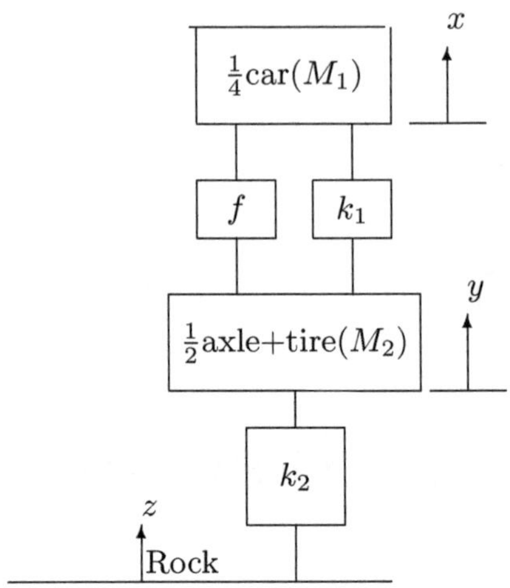

Figure 15.1: Diagram of Suspension System

15.2 Parameters

The parameters are M_1, M_2 f, k_1, k_2, z and the initial values. We will assume that the values of M_1, M_2, z and the initial values are all known precisely and will be crisp (not fuzzy). However spring constants and damping constants are always difficult to measure exactly so their values will be fuzzy. Employing expert advice (Chapter 3) we obtain triangular fuzzy numbers for k_1, k_2 and f and all the parameter values are given in Table 15.1. Using fuzzy parameters we get fuzzy differential equations and the solutions $\overline{x}(t)$ and $\overline{y}(t)$ are also fuzzy. The system now becomes a continuous fuzzy system whose trajectories are fuzzy so that any slice through a trajectory at some time t_z is a fuzzy number. We wish to estimate the band of uncertainty which is $\overline{x}(t)[0]$ and $\overline{y}(t)[0]$, or the $\alpha = 0$ cuts.

Next we need to choose the values of the fuzzy parameters in their $\alpha = 0$ cut to approximate the outer boundary of the $\alpha = 0$ cut of the fuzzy trajectory for the car and the tire. As discussed in Chapter 6 if $\overline{\omega} = (\omega_1/\omega_2/\omega_3)$ is a fuzzy parameter we use $\omega = \omega_1$, ω_2 and ω_3. We do this for $\overline{f}$, $\overline{k}_1$ and $\overline{k}_2$ giving $3^3 = 27$ curves for $x(t)$ and 27 graphs for $y(t)$ for a minimal approximation to $\overline{x}(t)[0]$ and $\overline{y}(t)[0]$. These are shown in Figures 15.3 and 15.4. The systems diagram for Simulink is in Figure 15.2.

In Figure 15.3 (15.4) the reader can determine, from consulting Section

Parameter	Fuzzy/Crisp Value
M_1	500 kg
M_1	20 kg
$\overline{f}$	(2000/2250/2500) Ns/m
$\overline{k}_1$	(7000/7500/8000) N/m
$\overline{k}_2$	(140000/150000/160000) N/m
z	0.05 m
$x(0)$	0
$y(0)$	z
$\dot{x}(0)$	0
$\dot{y}(0)$	0

Table 15.1: Fuzzy/Crisp Parameter Values in the Suspension System Example

1.5.4, $x(t)$ $(y(t))$ when we used for the fuzzy parameters: (1) all left end points of their $\alpha = 0$ cut; (2) all right end points of their $\alpha = 0$ cut; and (3) their $\alpha = 1$ values.

15.3 Simulation

Let us first explain the Simulink network in Figure 15.2. The top loop is for the car $(x(t))$ and the bottom loop is for the tire $(y(t))$. The input to "Integrate x'' " is $\ddot{x}$ from equation (15.1) with output $\dot{x}$. The output $\dot{x}$ goes to the box "$x' - y'$", whose output gets multiplied by constant f for input to "Add" and "Add1", and the output also travels to "Integrate x' " whose output is $x(t)$. $x(t)$ moves to the box "x-y" whose output is multiplied by the constant k_1 for input to "Add" and "Add1". $x(t)$ is also sent to "Scope Car" for its graph but "Car File" collects all the data on all the graphs to construct one figure containing all graphs. More information on how this is done is in Chapter 28. The output from "Add" ("Add1") is multiplied by $1/M_1$ $(1/M_2)$ for input to "Integrate x'' " ("Integrate y'' "). The bottom of the figure is for the input of the constant z. z is sent to the box "z-y" whose output gets multiplied by constant k_2 for input to "Add1". It should now be clear how we compute $y(t)$. There are also other items to set/choose, like type of numerical integrator and step sizes, which we will not discuss here (see Chapter 28).

Let $\overline{x}(t)[0] = [x_1(t,0), x_2(t,0)]$. In Figure 15.3 connect the upper boundary with a smooth curve which is our approximation to $x_2(t,0)$ and connect the lower boundary with a smooth curve producing an approximation to $x_1(t,0)$. The horizontal axis is time and the vertical axis is the displacement of the car from equilibrium. There is very little uncertainty since the

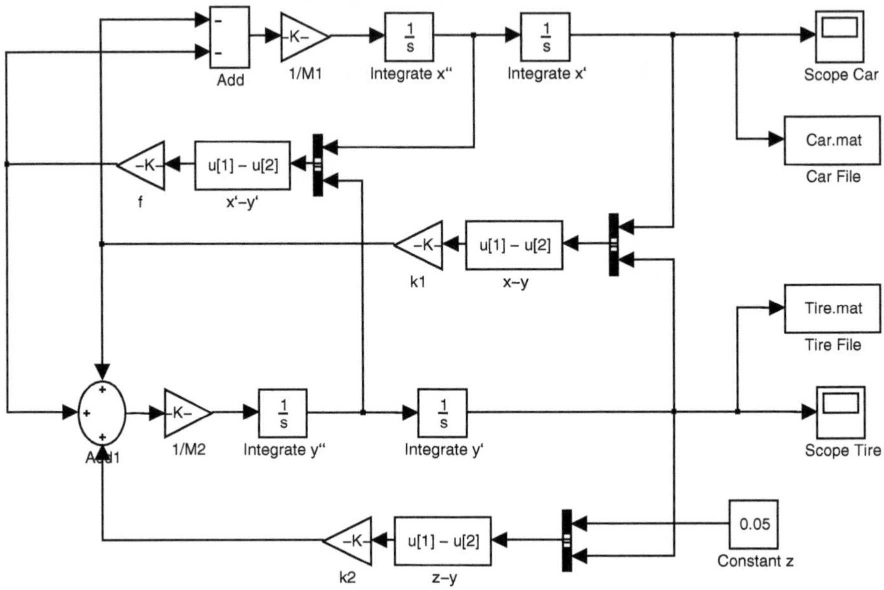

Figure 15.2: Simulink Diagram for the Suspension System Example

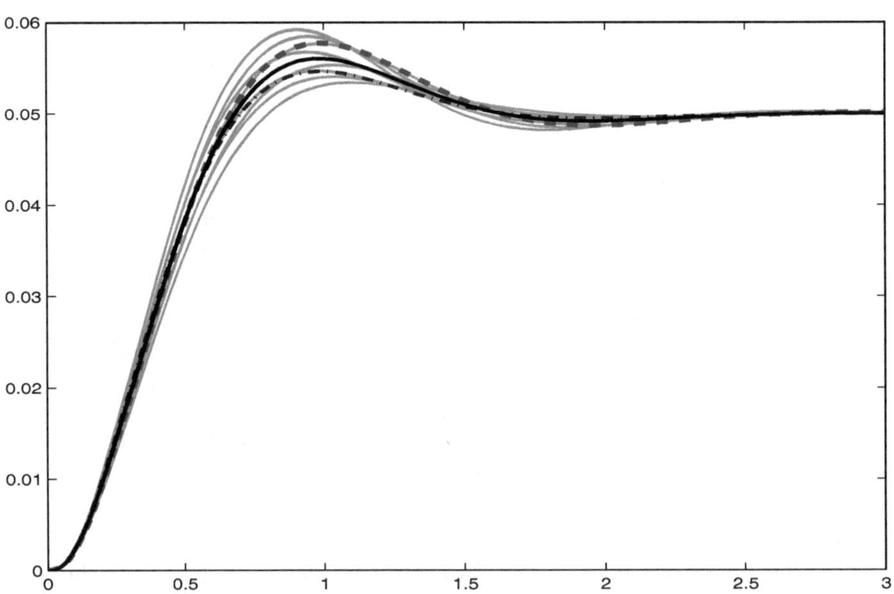

Figure 15.3: Fuzzy Trajectory for the Car $\overline{x}(t)[0]$ in the Suspension System Example

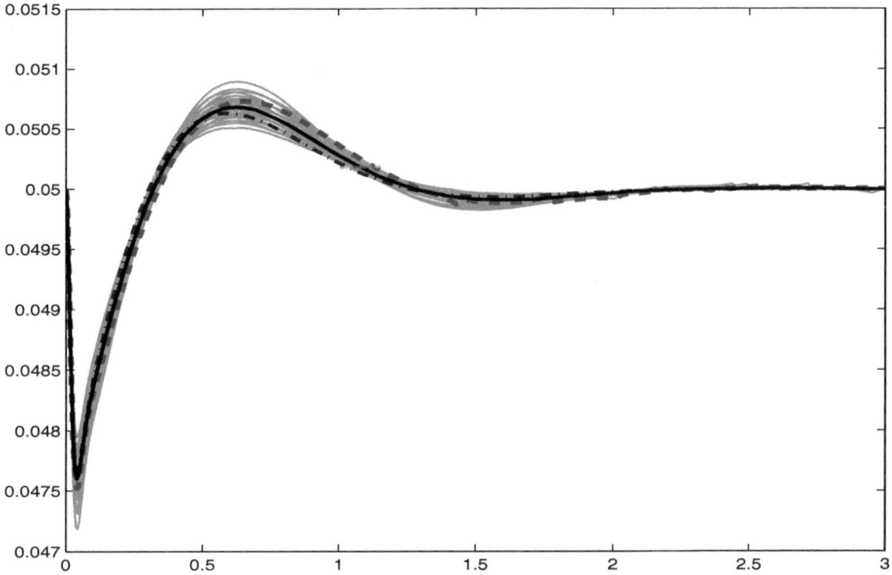

Figure 15.4: Fuzzy Trajectory for the Tire $\overline{y}(t)[0]$ in the Suspension System Example

uncertainty band is so narrow. The car gets displaced slightly more than the z value and then converges back to $z = 0.05$. The car shows very little oscillation, which shows the suspension system is doing a good job.

Let $\overline{y}(t)[0] = [y_1(t,0), y_2(t,0)]$. In Figure 15.4 connect the upper boundary with a smooth curve which is our approximation to $y_2(t,0)$ and connect the lower boundary with a smooth curve producing an approximation to $y_1(t,0)$. The horizontal axis is time and the vertical axis is the displacement of the tire from equilibrium. Again, there is very little uncertainty because the uncertainty band is extremely narrow. The tire very quickly oscillates from 0.05 to ≈ 0.0475 to ≈ 0.055 and then converges to the z value of 0.05. This suspension system is doing a very good job.

15.4 Reference

1. D.Matko, B.Zupancic and R.Karba: Simulation and Modelling of Continuous Systems, Prentice Hall, Englewood Cliffs, NJ, 1992.

Chapter 16

Chemical Reactions

16.1 Introduction

This application is about the Belousov-Zhabotinskii (BZ) chemical reaction ([1],[2]). It has to do with an oscillating chemical reaction discovered by Belousov in 1951 and the study of this reaction was continued by Zhabotinskii in 1964. It is an important result about chemical reactions and is considered the prototype of oscillating reactions. The system involves five chemicals. A system of nonlinear differential equations governing the evolution of the system for three of the chemicals is

$$\dot{x} = a_1 y - a_2 xy + a_3 x(1 - x), \tag{16.1}$$

$$\dot{y} = -b_1 y - b_2 xy + b_3 z, \tag{16.2}$$

$$\dot{z} = x - z, \tag{16.3}$$

where the a_i and the b_i are constants and the notation $\dot{x}$ ($\dot{y}$, $\dot{z}$) represents the time derivative of x (y,z). The initial conditions are $x(0) = x_0$, $y(0) = y_0$ and $z(0) = z_0$. All the constants are positive. $x(t)$ ($y(t)$, $z(t)$) is the concentration of one of the chemicals at time t. Two chemicals have been omitted because it is assumed that one of them has its concentration held fixed (so its derivative is zero) and the other chemical was considered not important and deleted from the model [1]. The derivation of these differential equations follows from: (1) the chemical models of the reactions (there are five of them); (2) the law of mass action [3]; and (3) the discussion in ([1],[2]). We will omit the details of the derivation.

16.2 Parameters

The constants a_i, $i = 1, 2, 3$ and b_1, b_2 will be assumed to be crisp and known. They depend on rate of reaction constants and the constant concentration

Parameter	Fuzzy/Crisp Value
a_1	0.5
a_2	100
a_3	100
b_1	0.015
b_2	3
b_3	$\overline{b}_3 = (1.4/1.8/2.2)$
x_0	0.5
y_0	0.1
z_0	0.4

Table 16.1: Fuzzy/Crisp Parameter Values in the Chemical Reaction Example

of one of the omitted chemicals, all assumed known. However, constant b_3 depends on a constant f, called a stoichiometric factor (dealing with the laws of definite proportions and of the conservation of mass and energy with respect to chemical activity), and its value is uncertain and important to the evolution of the system. Therefore, we will take f, and hence b_3, as a triangular fuzzy number obtained from expert opinion (Chapter 3). The values of all the parameters, including those for the initial conditions, are shown in Table 16.1.

Even with only one fuzzy parameter, the system becomes a system of fuzzy differential equations. The solutions will be fuzzy trajectories $\overline{x}(t)$, $\overline{y}(t)$ and $\overline{z}(t)$. A slice through a fuzzy trajectory at any fixed value of t gives a fuzzy number. We want to estimate the band of uncertainty which is the graph of the bases of these fuzzy numbers. That is, the graphs of $\overline{x}(t)[0]$, $\overline{y}(t)[0]$ and $\overline{z}(t)[0]$.

Next we need to choose the values of the fuzzy parameters in their $\alpha = 0$ cut to approximate the outer boundary of the $\alpha = 0$ cut of the fuzzy trajectory for $\overline{x}$, $\overline{y}$ and $\overline{z}$. As discussed in Chapter 6 if $\overline{\omega} = (\omega_1/\omega_2/\omega_3)$ is a fuzzy parameter we use $\omega = \omega_1$, ω_2 and ω_3. If we do this for $\overline{b}_3$ we will have only three curves because only one parameter is fuzzy. Since we have just b_3 fuzzy let us use the following spread of values in the $\alpha = 0$ cut: $b_3 = 1.4, 1.5, 1.6, 1.7, 1.8, 1.9, 2.0, 2.1, 2.2$. This will give us 9 curves for a minimal approximation to $\overline{x}(t)[0]$, $\overline{y}(t)[0]$ and $\overline{z}(t)[0]$. These are shown in Figures 16.2 - 16.4. The systems diagram for Simulink is in Figure 16.1.

In Figure 16.2 (16.3,16.4) the reader can determine, from consulting Section 1.5.4, $x(t)$ ($y(t)$,$z(t)$) when we used for the fuzzy parameter b_3: (1) the left end point of its $\alpha = 0$ cut; (2) the right end point of its $\alpha = 0$ cut; and (3) its $\alpha = 1$ value.

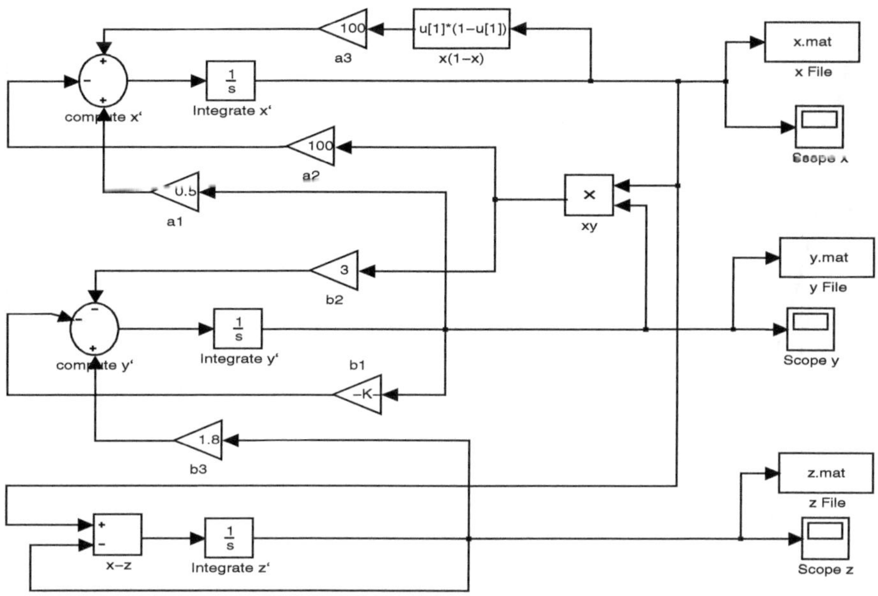

Figure 16.1: Simulink Diagram for the Chemical Reaction Example

16.3 Simulation

We first discuss the Simulink diagram in Figure 16.1. The top loop is for $x(t)$, the middle loop calculates $y(t)$ and the bottom loop gets $z(t)$. The input to "Integrate x' " is the right side of equation (16.1) with output $x(t)$. Now $x(t)$ goes to the box "$x(1-x)$" and "Gain a_3 " which computes $a_3x(1-x)$. $x(t)$ also travels to the box labeled "xy" and "Gain a_2" ("Gain b_2") to get a_2xy (b_2xy) for "compute x' " ("compute y' "). $x(t)$ is also sent to the box "x-z" for input to "Integrate z' " and to "Scope x" for its graph. However, the box labeled "x File" collects all the data on the graphs for one figure containing the nine graphs. More information on how this is done is in Chapter 28. It should now be clear how Simulink obtains $y(t)$. There are also other items to set/choose, like relative/absolute tolerance and type of numerical integrator, which we will not discuss here (see Chapter 28).

Let $\overline{x}(t)[0] = [x_1(t,0), x_2(t,0)]$. In Figure 16.2 connect the upper boundary with a smooth curve which is our approximation to $x_2(t,0)$ and connect the lower boundary with a smooth curve producing an approximation to $x_1(t,0)$. The horizontal axis is time and the vertical axis is the concentration of chemical $x(t)$. There appears to be two types of behavior: oscillatory and steady state. The horizontal graphs indicate the steady state result. So, uncertainty in the value of b_3 can produce dynamically different outcomes.

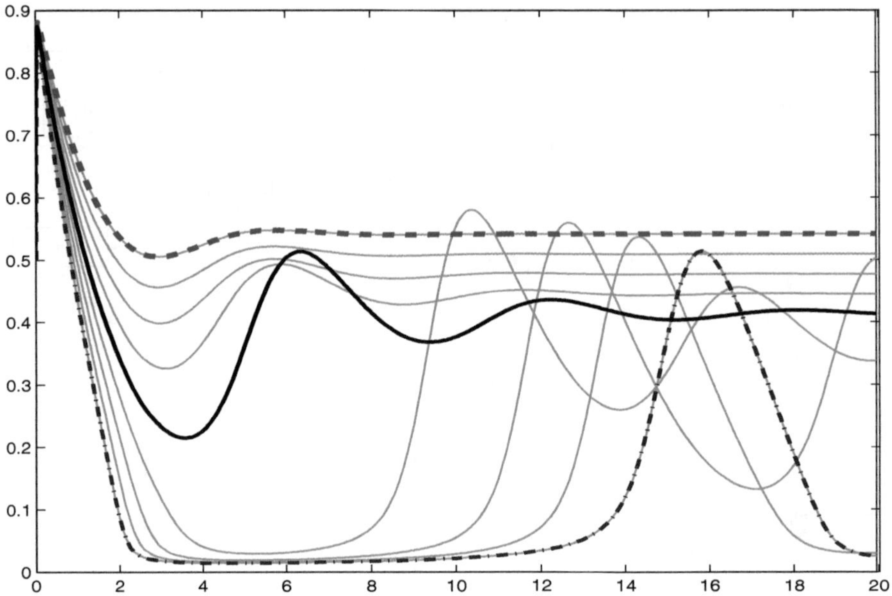

Figure 16.2: Fuzzy Trajectory for $\overline{x}(t)[0]$ in Chemical Reaction Example

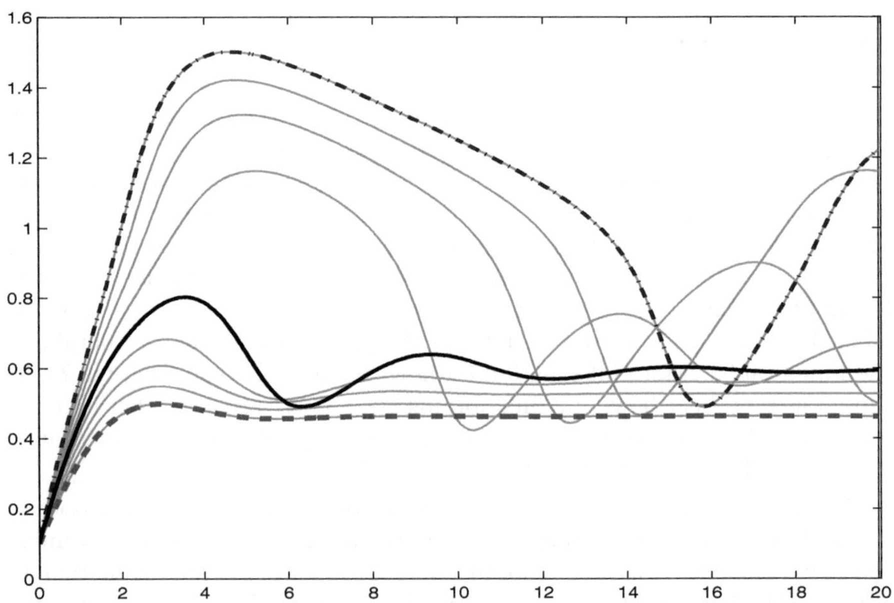

Figure 16.3: Fuzzy Trajectory for $\overline{y}(t)[0]$ in Chemical Reaction Example

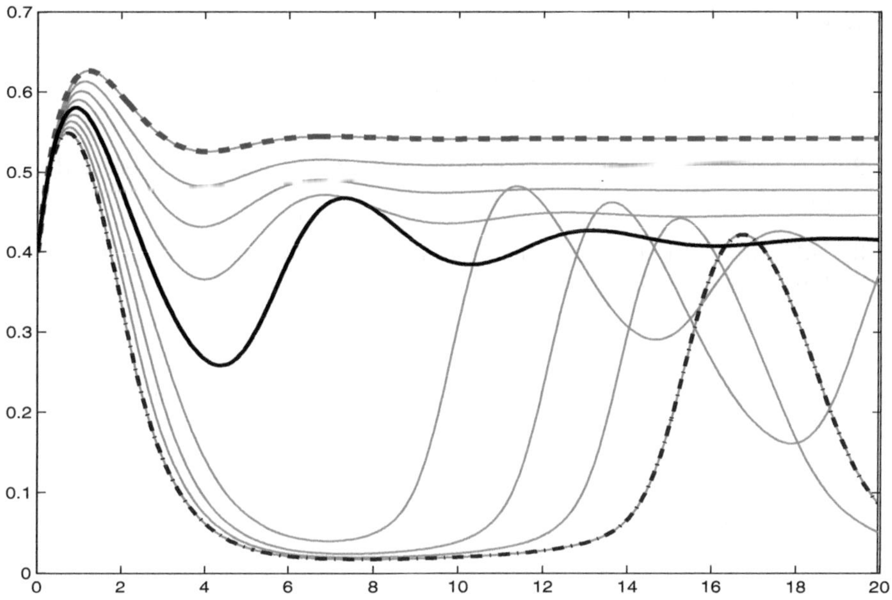

Figure 16.4: Fuzzy Trajectory for $\overline{z}(t)[0]$ in Chemical Reaction Example

Let $\overline{y}(t)[0] = [y_1(t,0), y_2(t,0)]$. In Figure 16.3 connect the upper boundary with a smooth curve which is our approximation to $y_2(t,0)$ and connect the lower boundary with a smooth curve producing an approximation to $y_1(t,0)$. The horizontal axis is time and the vertical axis is the concentration of chemical $y(t)$. Again, it looks like we can obtain oscillatory or steady state behavior depending on the value of b_3.

Let $\overline{z}(t)[0] = [z_1(t,0), z_2(t,0)]$. In Figure 16.4 connect the upper boundary with a smooth curve which is our approximation to $z_2(t,0)$ and connect the lower boundary with a smooth curve producing an approximation to $z_1(t,0)$. The horizontal axis is time and the vertical axis is the concentration of chemical $z(t)$. It is no surprise after Figures 16.2 and 16.3 that varying b_3 gives oscillation or steady state.

In practice in order to be able to predict the outcome we will need to have a more accurate estimate for b_3. We should point out that as $t \rightarrow \infty$ some of the oscillations die out and the graphs go horizontal but for $b_3 = 2.2$ the periodic motion of $x(t)$, $y(t)$ and $z(t)$ remain.

If all the uncertain parameters are estimated from data (Chapter 3), instead of using expert opinion, then the base of the fuzzy estimator is a 99% confidence interval. Using all of these confidence intervals we get $\overline{x}(t)[0]$, $\overline{y}(t)[0]$ and $\overline{z}(t)[0]$, so these bands of uncertainty will be like 99% confidence intervals for $x(t)$, $y(t)$ and $z(t)$, respectively.

16.4 References

1. J.D.Murry: Mathematical Biology, Springer-Verlag, Heidelberg, Germany, 1989.

2. S.H.Strogatz: Nonlinear Dynamics and Chaos, Addison-Wesley, Reading, MA., 1994.

3. D.G.Zill: A First Course in Differential Equations, Brooks/Cole, Pacific Grove, CA, 1997.

Chapter 17

The AIDS Epidemic

17.1 Introduction

This application is adapted from a model presented in [1]. The flow of people through the system is shown in Figure 17.1. The general population is all homosexual males in some geographical area (the US, western Europe, India,...). There is a constant immigration rate k_1 of susceptible males into the population of size $N(t)$. In this model the size of the population $N(t)$ can change in time, as compared to the Infectious Disease Model in Chapter 10, where the population size was fixed. In Figure 17.1: (1) $X(t)$ = number of susceptible males in the population; (2) $Y(t)$ = number of males infected with the HIV virus; (3) $A(t)$ = number of men with AIDS; and (4) $Z(t)$ = number infected with the virus but are non-infectious. The constants k_i, $i = 1, 2, ..., 5$ are all (positive) rates and time will be measured in years. People in X can go to ND = natural death with rate k_2, or to Y with rate k_3 times constant a = number of different sexual partners per year. Once in Y a man can go to ND with rate k_2, or to Z or A. Let b be the proportion of those infected with the virus who end up with AIDS. If k_4 is the rate for going from Y to A then we use bk_4 from Y to A and $(1-b)k_4$ from Y to Z in Figure 17.1. Once in A a person can go to ND at rate k_2 or die from AIDS DD at rate k_5. We may now present the system of linear differential equations for the system as

$$\dot{X} = k_1 - k_2X - ak_3X, \tag{17.1}$$

$$\dot{Y} = ak_3X - [k_4 + k_2]Y, \tag{17.2}$$

$$\dot{Z} = (1-b)k_4Y - k_2Z, \tag{17.3}$$

$$\dot{A} = bk_4Y - (k_2 + k_5)A, \tag{17.4}$$

$$N(t) = X(t) + Y(t) + Z(t) + A(t). \tag{17.5}$$

The initial conditions will be $A(0) = A_0 = 0$, $Z(0) = Z_0 = 0$, $X(0) + Y(0) = 100,000$ with $X(0) = X_0 = 90,000$, $Y(0) = Y_0 = 10,000$. We

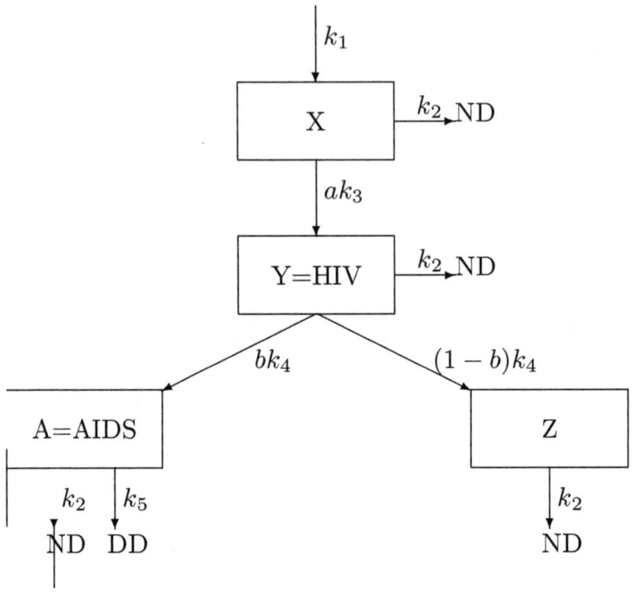

Figure 17.1: Diagram for the AIDS Epidemic

are primarily interested in the solutions for $A(t)$ and $Z(t)$. These are linear differential equations so we could easily find the solutions. However, even using simple solutions it can be complicated to find the boundary of the band of uncertainty, or the boundary of $\overline{A}(t)[0]$ and $\overline{Z}(t)[0]$, as shown in Section 6.2. So we omit expressions for the exact solutions and will use simulation instead.

17.2 Parameters

There are seven parameters k_i, $i = 1, 2, 3, 4, 5$, a and b in the model and all needed to be estimated from data. We assume that all the initial conditions are known and crisp. The fuzzy rates for $\overline{k}_i$, $i = 1, ..., 5$, might be estimated as in Section 3.4, $\overline{a}$ will be a fuzzy integer, and fuzzy b could be estimated as in Section 3.5. However, as in Chapter 6 we use three values for all the k_i, three values for b and five for a (see below), we would have $3^6 5 = 3645$ graphs for $A(t)$ and for $Z(t)$. Too many graphs to place on one coordinate system. So we must make a compromise. We will assume that all the rates k_i, $i = 1, ..., 5$ are known precisely and are crisp (not fuzzy), the initial conditions are also crisp but the two parameters a and b are to be estimated and are therefore fuzzy. Experts are asked to estimate a and b (Chapter 3) and we

Parameter	Fuzzy/Crisp Value
k_1	$13,333.30$
k_2	$\frac{1}{32}$
k_3	0.7
k_4	0.2
k_5	1.0
a	$\overline{a}$ = see equation (17.6)
b	$\overline{b} = (0.2/0.3/0.4)$
A_0	0
Z_0	0
X_0	$90,000$
Y_0	$10,000$

Table 17.1: Fuzzy/Crisp Parameter Values in the AIDS Epidemic

obtain a triangular fuzzy number for $\overline{b}$ but a fuzzy integer for $\overline{a}$. The results are given in Table 17.1. Fuzzy integers are defined like discrete fuzzy sets in Section 2.2.4. The fuzzy integer $\overline{a}$ in Table 17.1 is defined on the integers 2 through 6 as

$$\overline{a} = \{\frac{0.5}{2}, \frac{0.2}{3}, \frac{0.2}{4}, \frac{0.1}{5}, \frac{0.1}{6}\}, \tag{17.6}$$

which means that the possibility that $a = 2$ is 0.5, the possibility that $a = 3$ is 0.2, etc. The sum of the membership values need not equal one.

Using fuzzy parameters we get fuzzy differential equations and the solutions $\overline{A}(t)$ and $\overline{Z}(t)$ are also fuzzy. The system now becomes a continuous fuzzy system whose trajectories are fuzzy so that any slice through a trajectory at some time t_z is a fuzzy number. We wish to estimate the band of uncertainty which is $\overline{A}(t)[0]$ and $\overline{Z}(t)[0]$, or the $\alpha = 0$ cuts. We will also do this for $\overline{X}(t)[0]$ and $\overline{Y}(t)[0]$.

Next we need to choose the values of the fuzzy parameters in their $\alpha = 0$ cut to approximate the outer boundary of the $\alpha = 0$ cut of the fuzzy trajectory for $\overline{A}(t)$ and $\overline{Z}(t)$. As discussed in Chapter 6 if $\overline{\omega} = (\omega_1/\omega_2/\omega_3)$ is a fuzzy parameter we use $\omega = \omega_1$, ω_2 and ω_3. We do this for $\overline{b}$ but we use $a = 2, 3, 4, 5, 6$ giving 15 curves for $A(t)$ and $Z(t)$ for a minimal approximation to $\overline{A}(t)[0]$ and $\overline{Z}(t)[0]$. Now, from equations (17.1) and (17.2), $X(t)$ and $Y(t)$ depend only on fuzzy a so there will be five curves for $X(t)$ and $Y(t)$ for a minimal approximation to $\overline{X}(t)[0]$ and $\overline{Y}(t)[0]$. All of these graphs are shown in Figures 17.3-17.6. The systems diagram for Simulink is in Figure 17.2.

In Figure 17.3-17.6 the reader can determine, from consulting Section 1.5.4, $A(t)$ ($Z(t)$,$X(t)$,$Y(t)$) when we used for the fuzzy parameters: (1) all left end points of their $\alpha = 0$ cut; (2) all right end points of their $\alpha = 0$ cut; and (3) their $\alpha = 1$ values.

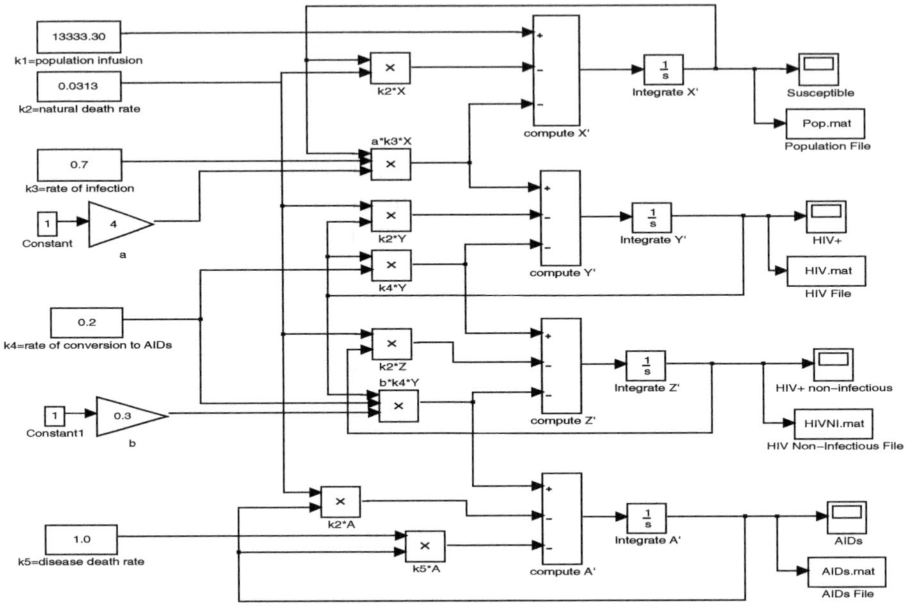

Figure 17.2: Simulink Diagram for the AIDS Epidemic

17.3 Simulation

Let us first discuss how Simulink simulated the problem. There are four loops: (1) the top loop is for $X(t)$; (2) the next loop is for $Y(t)$; (3) the third computes $Z(t)$; and (4) the last loop does $A(t)$. All loops are similar so we will only explain the top loop for $X(t)$. The input to "Integrate X'" is the right side of equation (17.1) with output $X(t)$. $X(t)$ travels to the box called "k_2X" where it computes this value and $X(t)$ goes to the box "ak_3X" which calculates its value. The input to "compute X'" is k_1, k_2X and ak_3X with output $k_1 - k_2X - ak_3X$. Also, $X(t)$ is sent to the "Scope" called "Susceptible" to show its graph but the box "Population File" gets the data on all the graphs to create one figure containing the graphs. More information on how this is done is in Chapter 28. There are also other items to set/choose, like simulation time and step size, which we will not discuss here (see Chapter 28).

Let $\overline{A}(t)[0] = [a_1(t,0), a_2(t,0)]$. In Figure 17.3 connect the upper boundary with a smooth curve which is our approximation to $a_2(t,0)$ and connect the lower boundary with a smooth curve producing an approximation to $a_1(t,0)$. The horizontal axis is time and the vertical axis is the number of men with AIDS. We get steady state (horizontal curves) with $\lim_{t\to\infty} a_2(t,0) \approx 4500$ and $\lim_{t\to\infty} a_1(t,0) \approx 2200$. The uncertainty band,

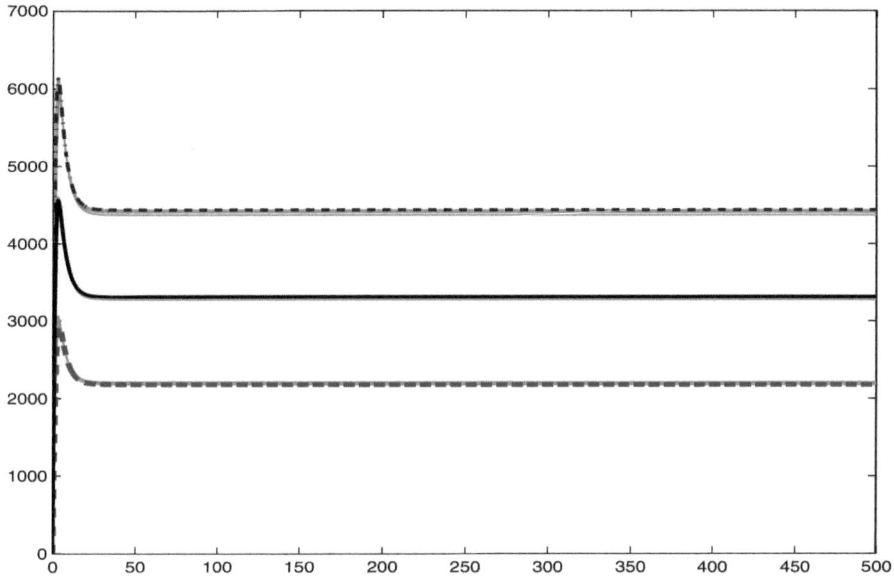

Figure 17.3: Fuzzy Trajectory for AIDS $\overline{A}(t)[0]$ in the AIDS Epidemic

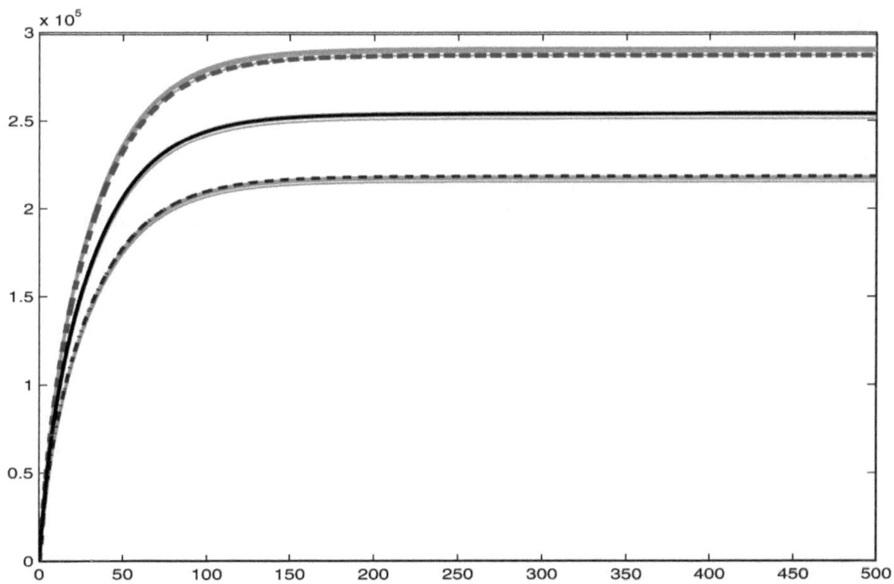

Figure 17.4: Fuzzy Trajectory for Non-Infectious $\overline{Z}(t)[0]$ in the AIDS Epidemic

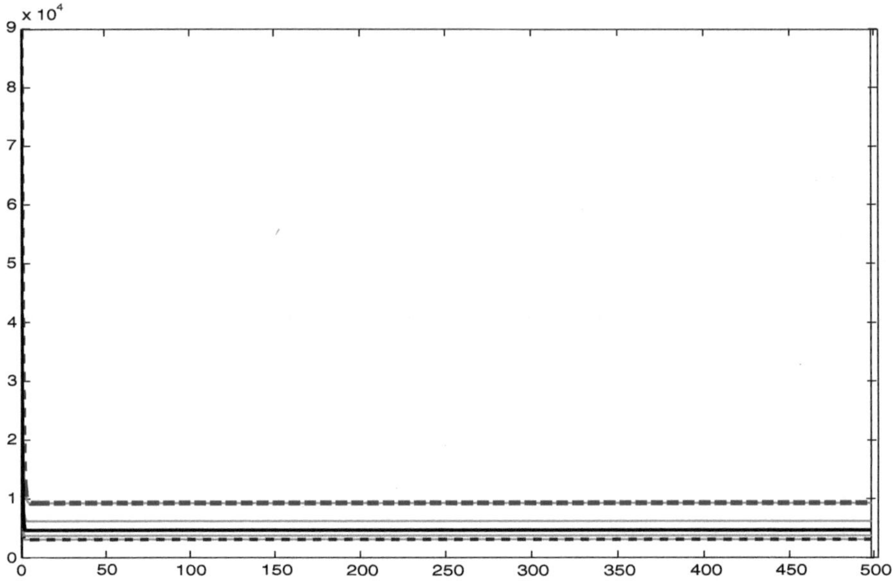

Figure 17.5: Fuzzy Trajectory for Susceptible $\overline{X}(t)[0]$ in the AIDS Epidemic

the difference between the two curves in steady state, is approximately 2300. This is a reasonable amount of uncertainty for those with AIDS. However, we see that the number of men with AIDS can quickly peak at approximately 6200.

Let $\overline{Z}(t)[0] = [z_1(t, 0), z_2(t, 0)]$. In Figure 17.4 connect the upper boundary with a smooth curve which is our approximation to $z_2(t, 0)$ and connect the lower boundary with a smooth curve producing an approximation to $z_1(t, 0)$. The horizontal axis is time and the vertical axis is the number of males infected with the virus but are non-infectious. We get steady state (horizontal curves) with $\lim_{t\to\infty} z_2(t, 0) \approx 280,000$ and $\lim_{t\to\infty} z_1(t, 0) \approx 220,000$. The uncertainty band, the difference between the two curves in steady state, is approximately $60,000$. This is a rather large amount of uncertainty of those which are HIV positive but non-infectious.

Let $\overline{X}(t)[0] = [x_1(t, 0), x_2(t, 0)]$. In Figure 17.5 connect the upper boundary with a smooth curve which is our approximation to $x_2(t, 0)$ and connect the lower boundary with a smooth curve producing an approximation to $x_1(t, 0)$. The horizontal axis is time and the vertical axis is the number of susceptible males in the population. We see that $X(t)$ rapidly decreases to its steady state values. We get steady state (horizontal curves) with $\lim_{t\to\infty} x_2(t, 0) \approx 10,000$ and $\lim_{t\to\infty} x_1(t, 0) \approx 4,000$. The uncertainty band, the difference between the two curves in steady state, is approximately $6,000$. This is a moderately large amount of uncertainty for those which are susceptible.

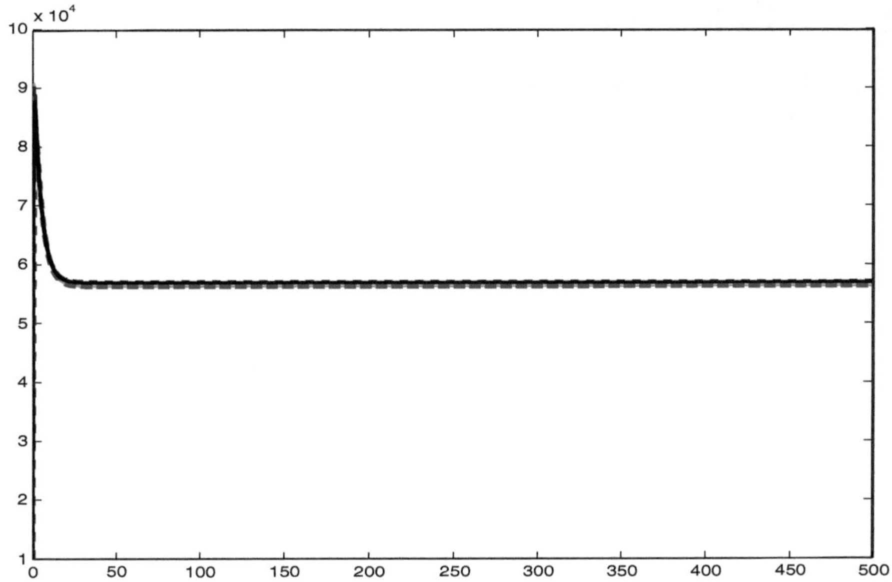

Figure 17.6: Fuzzy Trajectory for HIV $\overline{Y}(t)[0]$ in the AIDS Epidemic

Let $\overline{Y}(t)[0] = [y_1(t,0), y_2(t,0)]$. In Figure 17.6 connect the upper boundary with a smooth curve which is our approximation to $y_2(t,0)$ and connect the lower boundary with a smooth curve producing an approximation to $y_1(t,0)$. The horizontal axis is time and the vertical axis is the number of males infected with the HIV virus. We get steady state (horizontal curves) with $\lim_{t\to\infty} y_2(t,0) = \lim_{t\to\infty} y_1(t,0) \approx 57,000$. The uncertainty band, the difference between the two curves in steady state, is approximately zero. However, $Y(t)$ may rapidly increase to approximately $90,000$ before approaching its steady state value.

There was not a lot of uncertainty imputed to this model. We only have two fuzzy parameters a and b. We essentially obtain three curves in Figures 17.3, 17.4 and 17.5 corresponding to the three values for parameter b.

If all the uncertain parameters are estimated from data (Chapter 3), and none from expert opinion, then the base of the fuzzy estimator is a 99% confidence interval. Using all of these confidence intervals we get $\overline{A}(t)[0]$, $\overline{Z}(t)[0]$, $\overline{X}(t)[0]$ and $\overline{Y}(t)[0]$, so these bands of uncertainty will be like 99% confidence intervals for $A(t)$, $Z(t)$, $X(t)$ and $Y(t)$, respectively.

17.4 Reference

1. J.D.Murry: Mathematical Biology, Springer-Verlag, Heidelberg, Germany, 1989.

Chapter 18

The Machine/Service Queuing Model

18.1 Introduction

In this chapter we consider the classical M machine and R service person problem from queuing theory [4]. In this model we assume there are R repair persons available to service M machines. The calling source for the model, all the customers who can arrive at the queuing system, are the M machines which "arrive" for service when they break down. However, we do not bring a broken machine to the service station, a service station consisting of one service person travels to the broken machine to repair it. Assume that if a machine is operating correctly, the time to break down is described by an exponential distribution that has rate λ. Assume also that any machine that has broken down requires the service of a single repair person and that the service time probability distribution is exponential with the same rate μ for each repair person and each machine.

Usually in queuing theory we investigate the behavior of the system during "steady state". Let $p_i(t)$ be the probability of i broken machines at time $t \geq 0$, for $0 \leq i \leq M$. If $p_i = \lim_{t\to\infty} p_i(t)$, $0 \leq i \leq M$, then the p_i are called the steady state probabilities and using the p_i in calculations produces steady state results. However, in this chapter we do not go directly to the steady state probabilities but start with the differential equations defining the $p_i(t)$.

To simplify the discussion now assume that $M = 4$ and $R = 2$. We write $\dot{p}_i$ for the time derivative of $p_i(t)$. The following system of linear differential equations has been adopted from a discussion in [5].

$$\dot{p}_0 = -4\lambda p_0(t) + \mu p_1(t), \tag{18.1}$$

$$\dot{p}_1 = 4\lambda p_0(t) - [3\lambda + \mu]p_1(t) + 2\mu p_2(t), \tag{18.2}$$

Parameter	Fuzzy/Crisp Value
λ	$\overline{\lambda} = (0.3/0.5/0.7)$
μ	$\overline{\mu} = (1.5/2.0/2.5)$
$p_0(0)$	1.00
$p_1(0)$	0
$p_2(0)$	0
$p_3(0)$	0
$p_4(0)$	0

Table 18.1: Fuzzy/Crisp Parameters in Machine/Service Queuing Model

$$\dot{p}_2 = 3\lambda p_1(t) - [2\lambda + 2\mu]p_2(t) + 2\mu p_3(t), \tag{18.3}$$

$$\dot{p}_3 = 2\lambda p_2(t) - [\lambda + 2\mu]p_3(t) + 2\mu p_4(t), \tag{18.4}$$

$$\dot{p}_4 = \lambda p_3(t) - 2\mu p_4(t). \tag{18.5}$$

Initial conditions are $p_0(0) = 1$ and $p_1(0) = p_2(0) = p_3(0) = p_4(0) = 0$. The solution to this system produces the $p_i(t)$, $0 \leq i \leq 4$. We wish to look at $p_0(t)$ and we will also be interested in finding $N(t) = p_1(t)+2p_2(t)+3p_3(t)+4p_4(t)$ the expected number of broken machines in the system at time t.

These are linear differential equations so we could easily find the solutions. However, even using simple solutions it can be complicated to find the boundary of the band of uncertainty, or the boundary of $\overline{p}_0(t)[0]$ and $\overline{N}(t)[0]$, as shown in Section 6.2. So we omit expressions for the exact solutions and will use simulation instead.

18.2 Parameters

The parameters λ and μ need to be estimated from data. So, from ([1]-[3], see also Section 3.4) their estimators become fuzzy numbers which exhibit the uncertainty in their values. Let us now use $\overline{\lambda} = (0.3/0.5/0.7)$ as our fuzzy estimator of λ, or approximately one breakdown per machine every two hours. Time is measured in hours. And let $\overline{\mu} = (1.5/2/2.5)$ be our fuzzy estimator of μ, or approximately 30 minutes per repair. The values of all the parameters is shown in Table 18.1.

Using fuzzy parameters we get fuzzy differential equations and the solutions $\overline{p}_i(t)$, $i = 0, ..., 4$, and $\overline{N}(t)$ are also fuzzy. The system now becomes a continuous fuzzy system whose trajectories are fuzzy so that any slice through a trajectory at some time t_z is a fuzzy number. We wish to estimate certain bands of uncertainty which are $\overline{p}_0(t)[0]$ and $\overline{N}(t)[0]$, or the $\alpha = 0$ cuts.

Next we need to choose the values of the fuzzy parameters in their $\alpha = 0$ cut to approximate the outer boundary of the $\alpha = 0$ cut of the fuzzy trajectory for $\overline{p}_0(t)$ and $\overline{N}(t)$. As discussed in Chapter 6 if $\overline{\omega} = (\omega_1/\omega_2/\omega_3)$

is a fuzzy parameter we use $\omega = \omega_1$, ω_2 and ω_3. We do this for $\overline{\lambda}$ and $\overline{\mu}$ giving $3^2 = 9$ curves for $p_0(t)$ and 9 graphs for $N(t)$ for a minimal approximation to $\overline{p}_0(t)[0]$ and $\overline{N}(t)[0]$. These are shown in Figures 18.2-18.3 and in Figures 18.4-18.5. The systems diagram for Simulink is in Figure 18.1.

In Figures 18.2-18.5 the reader can determine, from consulting Section 1.5.4, $p_0(t)$ ($N(t)$) when we used for the fuzzy parameters: (1) all left end points of their $\alpha = 0$ cut; (2) all right end points of their $\alpha = 0$ cut; and (3) their $\alpha = 1$ values.

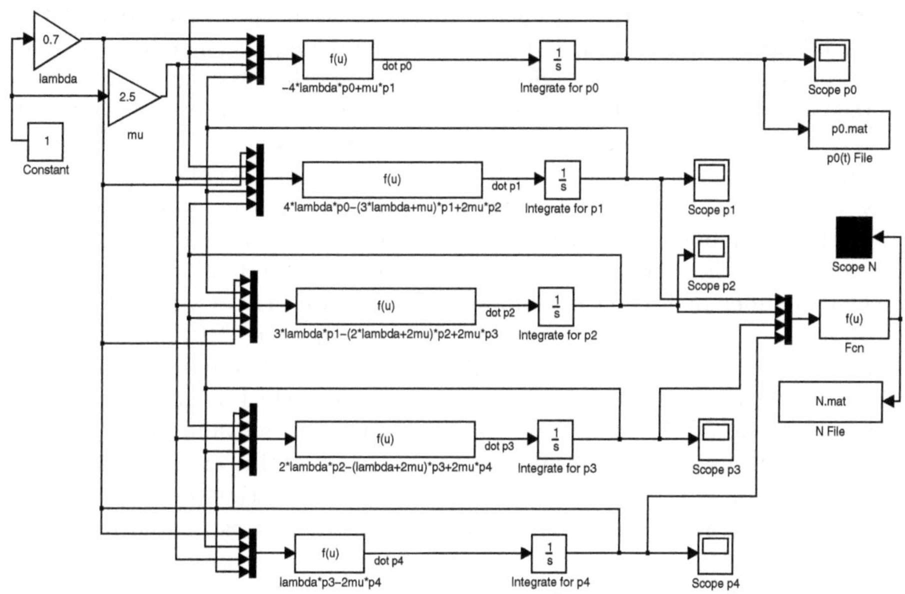

Figure 18.1: Simulink Diagram for the Machine/Service Queuing Model

18.3 First Simulation

We first describe Figure 18.1. There are five loops, with the top for $p_0(t)$ and the bottom for $p_4(t)$. Let us only discuss the top loop because the others are very similar. The input to "Integrate for p_0" is the right side of equation (18.1) and its output is $p_0(t)$. Now $p_0(t)$ cycles around the top for input to the top box labeled "f(u)". This box computes the right side of equation (18.1). $p_0(t)$ also is sent to "Scope p_0" to show its graph. The box "$p_0(t)$ File" collects all the data to construct one figure with all nine graphs. Chapter 28 has more details on how this is accomplished. The other boxes labeled "f(u)" calculate the right hand side of equations (18.2)-(18.5). In the middle of the right side, the box "f(u)" determines $N(t)$ from the $p_i(t)$, $i = 0, ..., 4$, for input to "Scope N" (its graph) and "N File" (for all graphs). There are also other items to set/choose, like initial step size and type of numerical integrator (fixed or variable steps), which we will not discuss here (see Chapter 28).

Let $\overline{p}_0(t)[0] = [p_{01}(t,0), p_{02}(t,0)]$. In Figure 18.2 connect the upper boundary with a smooth curve which is our approximation to $p_{02}(t,0)$ and connect the lower boundary with a smooth curve producing an approximation to $p_{01}(t,0)$. The horizontal axis is time and the vertical axis is the probability of no broken machines. Notice how fast the system gets into steady state. Steady state arrives when the curves go horizontal. The steady state probabilities are $\lim_{t\to\infty} p_{02}(t,0) = 0.6340$ and $\lim_{t\to\infty} p_{01}(t,0) = 0.2030$. The uncertainty band, the difference between both curves at steady state, is $0.6340 - 0.2030 = 0.4310$ which is quite large. This uncertainty on $p_0(t)$ results from the uncertainty in the values for λ and μ.

Let $\overline{N}(t)[0] = [n_1(t,0), n_2(t,0)]$. In Figure 18.3 connect the upper boundary with a smooth curve which is our approximation to $n_2(t,0)$ and connect the lower boundary with a smooth curve producing an approximation to $n_1(t,0)$. The horizontal axis is time and the vertical axis is the expected number of broken machines. Steady state arrives when the curves go horizontal. The steady state values are $\lim_{t\to\infty} n_2(t,0) = 1.3965$ and $\lim_{t\to\infty} n_1(t,0) = 0.4353$. The uncertainty band, the difference between both curves at steady state, is $1.3965 - 0.4353 = 0.9612$ which is quite large. This uncertainty on $N(t)$ results from the uncertainty in the values for λ and μ.

Since the uncertainty results are quite large in this example we would concentrate on obtaining more accurate (less fuzzy) estimates for λ and μ.

Notice that both uncertain parameters (λ, μ) were estimated from data and the base of their fuzzy estimator is a 99% confidence interval. So we can say that the bands of uncertainty are like bands of 99% confidence intervals. The same can be said for the other simulations discussed below.

18.4 Second Simulation

The fuzziness (width of the base of the fuzzy number) of the fuzzy number estimators for λ and μ is a function of m = the sample size (Section 3.4). So we gather more data on our system, get a larger value for m, to obtain less fuzzy estimators. Assume the new fuzzy estimators are the triangular fuzzy numbers $\overline{\lambda} = (0.4/0.5/0.6)$ and $\overline{\mu} = (1.8/2/2.2)$. We perform the simulation again with no other changes. The results are in Figures 18.4 and 18.5.

In Figure 18.4 connect the upper boundary with a smooth curve which is our approximation to $p_{02}(t,0)$ and connect the lower boundary with a smooth curve producing an approximation to $p_{01}(t,0)$. The horizontal axis is time and the vertical axis is the probability of no broken machines. Steady state arrives when the curves go horizontal. The steady state probabilities are $\lim_{t\to\infty} p_{02}(t,0) = 0.5089$ and $\lim_{t\to\infty} p_{01}(t,0) = 0.3068$. The uncertainty

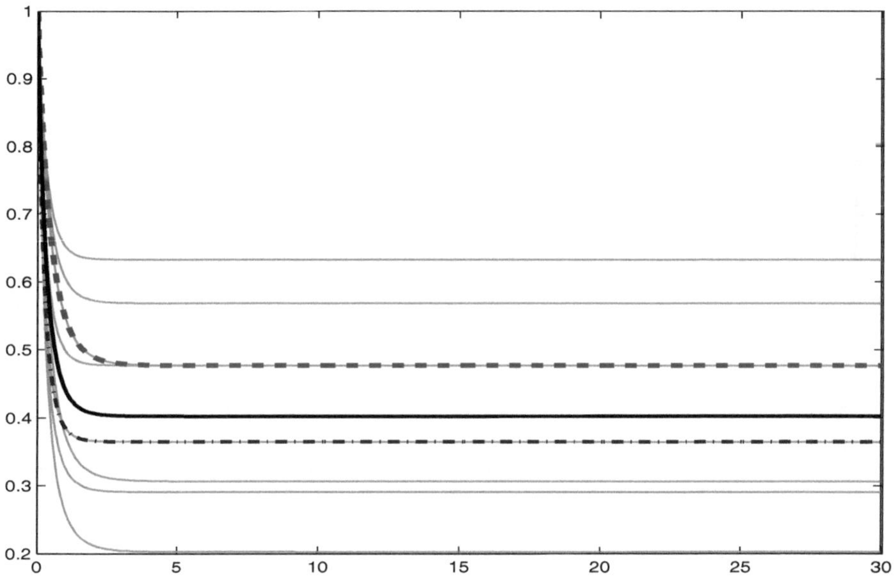

Figure 18.2: Fuzzy Trajectory of Fuzzy Probability $\overline{p}_0(t)[0]$ in the First Simulation

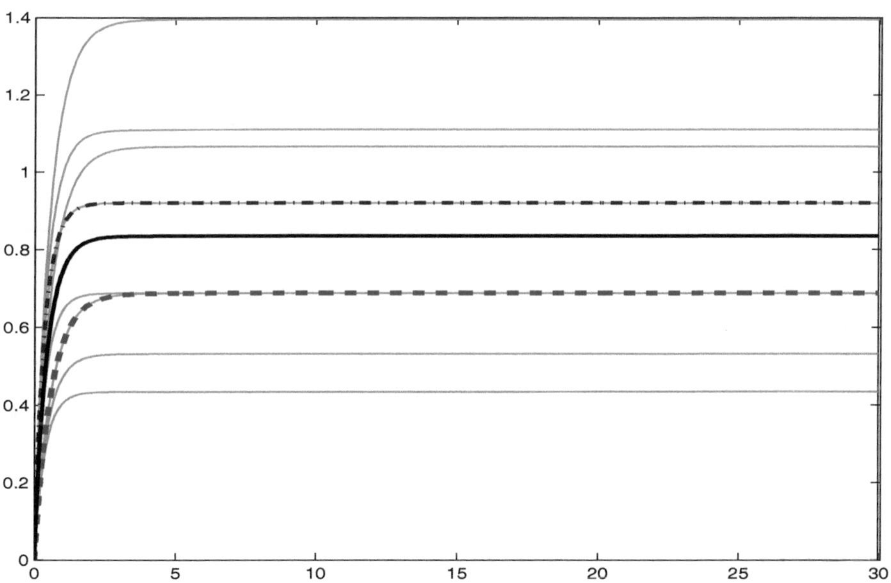

Figure 18.3: Fuzzy Trajectory of the Expected Number of Broken Machines $\overline{N}(t)[0]$ in the First Simulation

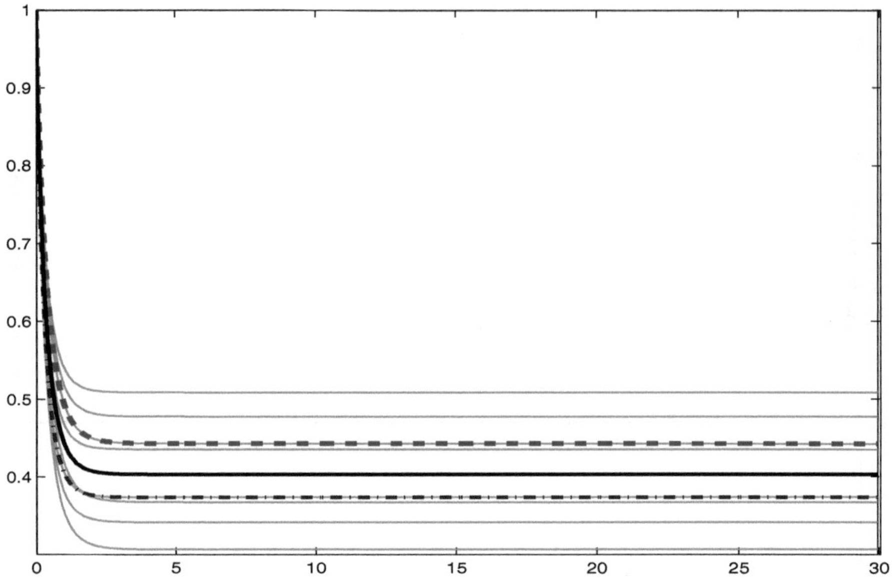

Figure 18.4: Fuzzy Trajectory of Fuzzy Probability $\overline{p}_0(t)[0]$ in the Second Simulation

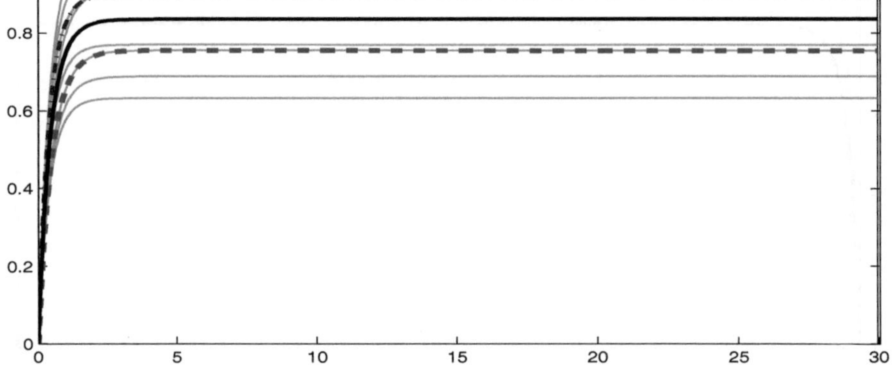

Figure 18.5: Fuzzy Trajectory of the Expected Number of Broken Machines $\overline{N}(t)[0]$ in the Second Simulation

band, the difference between both curves at steady state, is 0.2021 which is much smaller than that in the first simulation. This uncertainty on $p_0(t)$ results from the uncertainty in the values for λ and μ.

In Figure 18.5 connect the upper boundary with a smooth curve which is our approximation to $n_2(t,0)$ and connect the lower boundary with a smooth curve producing an approximation to $n_1(t,0)$. The horizontal axis is time and the vertical axis is the expected number of broken machines. Steady state arrives when the curves go horizontal. The steady state values are $\lim_{t\to\infty} n_2(t,0) = 1.0681$ and $\lim_{t\to\infty} n_1(t,0) = 0.6336$. The uncertainty band, the difference between both curves at steady state, is 0.4345 which is a lot smaller than that in the first simulation. This uncertainty on $N(t)$ results from the uncertainty in the values for λ and μ.

It is obvious that as the fuzziness in the fuzzy estimators decreases, the uncertainty bands in the fuzzy results also are reduced.

18.5 References

1. J.J.Buckley: Fuzzy Statistics, Springer, Heidelberg, Germany, 2004.

2. J.J.Buckley: Fuzzy Probabilities and Fuzzy Sets for Web Planning, Springer, Heidelberg, Germany, 2004.

3. J.J.Buckley: Simulating Fuzzy Systems, Springer, Heidelberg, Germany, 2005.

4. H.A.Taha: Operations Research, Fifth Edition, Macmillan, N.Y., 1992.

5. H.M.Wagner: Principles of Operations Research, Second Edition, Prentice Hall, Englewood Cliffs, N.J.. 1975.

Chapter 19

A Self-Service Queuing Model

19.1 Introduction

In this chapter we consider a queuing model with c identical and parallel servers and system capacity M with $M = c$ [4]. This means that there is no queue and an arriving customer enters the system only when there is an empty server; otherwise, if all servers are busy they leave the system. This is sometimes called the self-service model but usually such a model assumes there is an infinite number of servers. We will have a finite number of servers. So, if a customer arrives and a server is free, the customer enters the server and they "service" themselves.

Assume that the calling source (where the customers come from) is infinite and arrival time is described by an exponential distribution that has rate λ. Assume also that the service time probability distribution is exponential with the same rate μ for each server and each customer.

Usually in queuing theory we investigate the behavior of the system during "steady state". Let $p_i(t)$ be the probability of i customers in the system at time $t > 0$, for $0 \leq i \leq M$. If $p_i = \lim_{t\to\infty} p_i(t)$, $0 \leq i \leq M$, then the p_i are called the steady state probabilities and using the p_i in calculations produces steady state results. However, in this chapter we do not go directly to the steady state probabilities but start with the differential equations defining the $p_i(t)$.

To simplify the discussion now assume that $M = c = 4$. We write $\dot{p}_i$ for the time derivative of $p_i(t)$. The following system of linear differential equations has been adopted from a discussion in [5].

$$\dot{p}_0 = -\lambda p_0(t) + \mu p_1(t), \tag{19.1}$$

$$\dot{p}_1 = \lambda p_0(t) - [\lambda + \mu] p_1(t) + 2\mu p_2(t), \tag{19.2}$$

Parameter	Fuzzy/Crisp Value
λ	$\overline{\lambda} = (0.3/0.5/0.7)$
μ	$\overline{\mu} = (0.5/1.0/1.5)$
$p_0(0)$	1.00
$p_1(0)$	0
$p_2(0)$	0
$p_3(0)$	0
$p_4(0)$	0

Table 19.1: Fuzzy/Crisp Parameters in the Self-Service Queuing Model

$$\dot{p}_2 = \lambda p_1(t) - [\lambda + 2\mu]p_2(t) + 3\mu p_3(t), \tag{19.3}$$

$$\dot{p}_3 = \lambda p_2(t) - [\lambda + 3\mu]p_3(t) + 4\mu p_4(t), \tag{19.4}$$

$$\dot{p}_4 = \lambda p_3(t) - 4\mu p_4(t). \tag{19.5}$$

Initial conditions are $p_0(0) = 1$ and $p_1(0) = p_2(0) = p_3(0) = p_4(0) = 0$. The solution to this system produces the $p_i(t)$, $0 \leq i \leq 4$. We will be interested in finding $N(t) = p_1(t) + 2p_2(t) + 3p_3(t) + 4p_4(t)$ the expected number of customers in the system at time t and $p_4(t)100 = \%$ of lost customers due to capacity $M = c = 4$.

These are differential equations so we could easily find the solutions. However, even using simple solutions it can be complicated to find the boundary of the band of uncertainty, or the boundary of $\overline{p}_4(t)[0]$ and $\overline{N}(t)[0]$, as shown in Section 6.2. So we omit expressions for the exact solutions and will use simulation instead.

19.2 Parameters

The parameters λ and μ need to be estimated from data. So, from ([1]-[3], see also Section 3.4) their estimators become fuzzy numbers which exhibit the uncertainty in their values. Let us now use $\overline{\lambda} = (0.3/0.5/0.7)$ as our fuzzy estimator of λ, or approximately one customer arrival every two hours. And let $\overline{\mu} = (0.5/1/1.5)$ our fuzzy estimator of μ, or approximately one hour per service. Time is in hours. The values of all the parameters are given in Table 19.1.

Using fuzzy parameters we get differential equations and the solutions $\overline{p}_i(t)$, $i = 0, ..., 4$, and $\overline{N}(t)$ are also fuzzy. The system now becomes a continuous fuzzy system whose trajectories are fuzzy so that any slice through a trajectory at some time t_z is a fuzzy number. We wish to estimate certain bands of uncertainty which are $\overline{p}_4(t)[0]$ and $\overline{N}(t)[0]$, or the $\alpha = 0$ cuts.

Next we need to choose the values of the fuzzy parameters in their $\alpha = 0$ cut to approximate the outer boundary of the $\alpha = 0$ cut of the fuzzy trajectory for $\overline{p}_4(t)$ and $\overline{N}(t)$. As discussed in Chapter 6 if $\overline{\omega} = (\omega_1/\omega_2/\omega_3)$ is a fuzzy parameter we use $\omega = \omega_1, \omega_2$ and ω_3. We do this for $\overline{\lambda}$ and $\overline{\mu}$ giving $3^2 = 9$ curves for $p_4(t)$ and 9 graphs for $N(t)$ for a minimal approximation to $\overline{p}_4(t)[0]$ and $\overline{N}(t)[0]$. These are shown in Figures 19.2-19.3. The systems diagram for Simulink is in Figure 19.1.

In Figure 19.2 (19.3) the reader can determine, from consulting Section 1.5.4, $p_4(t)$ ($N(t)$) when we used for the fuzzy parameters: (1) all left end points of their $\alpha = 0$ cut; (2) all right end points of their $\alpha = 0$ cut; and (3) their $\alpha = 1$ values.

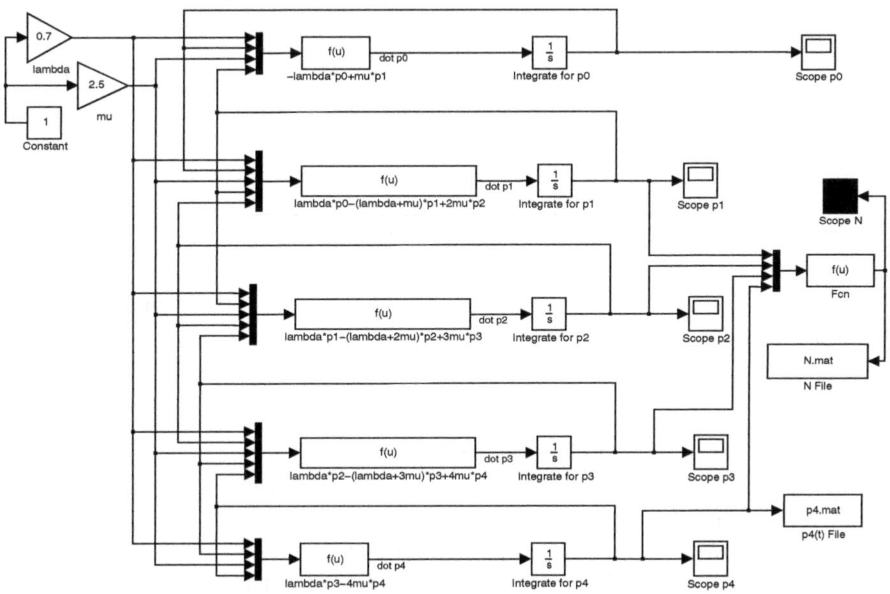

Figure 19.1: Simulink Diagram for the Self-Service Queuing Model

19.3 Simulation

The system diagram from Simulink in this application is similar to that in Figure 18.1 so we will be brief in our discussion of Figure 19.1. The five loops are to compute $p_i(t)$, $i = 0, ..., 4$. The "f(u)" boxes in the middle determine the right side of equation (19.1)-(19.5). The "f(u)" box on the right side computes $N(t)$ from the $p_i(t)$, $i = 0, ..., 4$. The "$p_4(t)$ File" and the "N File" collect all the data so that we may present all the graphs in one figure. Chapter 28 has more details on this. There are also other items to set/choose, like simulation time and type of numerical integrator, which we

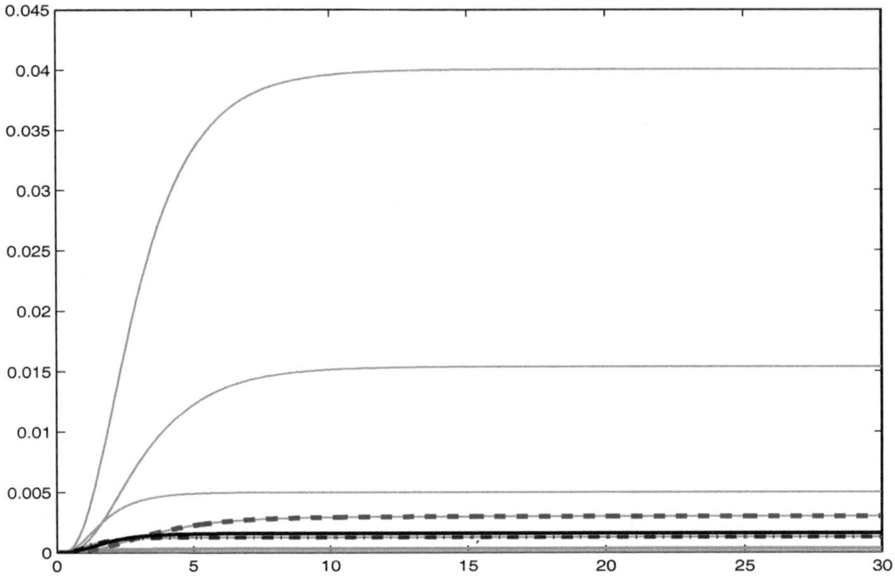

Figure 19.2: Fuzzy Trajectory of Fuzzy Probability $\overline{p}_4(t)[0]$ in the Self-Service Queuing Model

will not discuss here (see Chapter 28).

Let $\overline{p}_4(t)[0] = [p_{41}(t,0), p_{42}(t,0)]$. In Figure 19.2 connect the upper boundary with a smooth curve which is our approximation to $p_{42}(t,0)$ and connect the lower boundary with a smooth curve producing an approximation to $p_{41}(t,0)$. The horizontal axis is time and the vertical axis is the percent of lost customers (after multiplying by 100). Notice how fast the system gets into steady state. Steady state arrives when the curves go horizontal. The steady state probabilities are $\lim_{t\to\infty} p_{42}(t,0) = 0.0400$ and $\lim_{t\to\infty} p_{41}(t,0) = 0.0001$. The uncertainty band, the difference between both curves at steady state, is 0.0399 which is quite small. This uncertainty on $p_4(t)$ results from the uncertainty in the values for λ and μ.

Let $\overline{N}(t)[0] = [n_1(t,0), n_2(t,0)]$. In Figure 19.3 connect the upper boundary with a smooth curve which is our approximation to $n_2(t,0)$ and connect the lower boundary with a smooth curve producing an approximation to $n_1(t,0)$. The horizontal axis is time and the vertical axis is the expected number of customers in the system. Steady state arrives when the curves go horizontal. The steady state values are $\lim_{t\to\infty} n_2(t,0) = 1.3438$ and $\lim_{t\to\infty} n_1(t,0) = 0.2002$. The uncertainty band, the difference between both curves at steady state, is 1.1436 which is quite large. This uncertainty on $N(t)$ results from the uncertainty in the values for λ and μ.

Since the uncertainty results for $N(t)$ are quite large in this example we

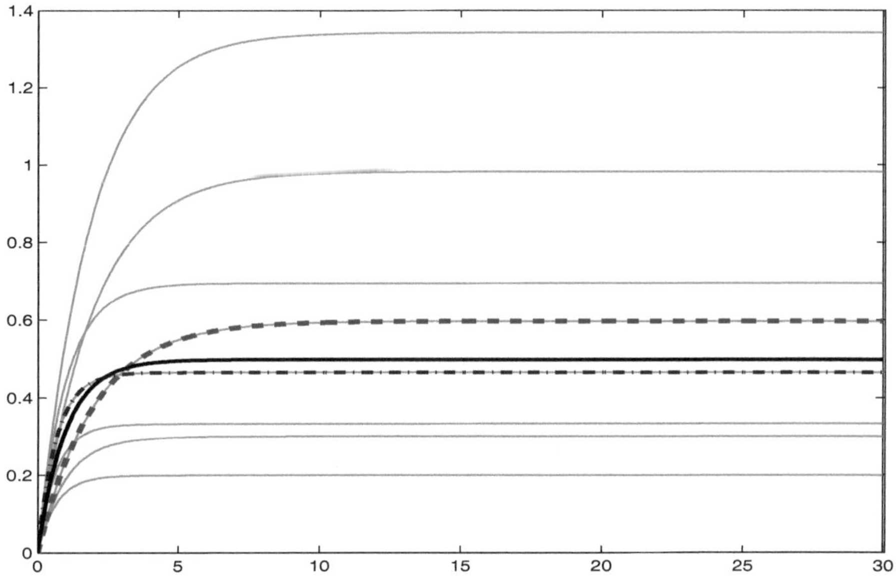

Figure 19.3: Fuzzy Trajectory of the Expected Number of Customers in the System $\overline{N}(t)[0]$ in the Self-Service Queuing Model

would concentrate on obtaining more accurate (less fuzzy) estimates for λ and μ as was done in the previous chapter.

Notice that both uncertain parameters (λ, μ) were estimated from data and the base of their fuzzy estimator is a 99% confidence interval. So we can say that the bands of uncertainty are like bands of 99% confidence intervals.

19.4 References

1. J.J.Buckley: Fuzzy Statistics, Springer, Heidelberg, Germany, 2004.

2. J.J.Buckley: Fuzzy Probabilities and Fuzzy Sets for Web Planning, Springer, Heidelberg, Germany, 2004.

3. J.J.Buckley: Simulating Fuzzy Systems, Springer, Heidelberg, Germany, 2005.

4. H.A.Taha: Operations Research, Fifth Edition, Macmillan, N.Y., 1992.

5. H.M.Wagner: Principles of Operations Research, Second Edition, Prentice Hall, Englewood Cliffs, N.J.. 1975.

Chapter 20

Symbiosis

20.1 Introduction

There are examples where the interaction of two species is to the benefit of both (symbiosis). Examples are plant and seed dispersers, plants and pollen dispersers, etc. A simple two species model, with limited carrying capacities for both species, adapted from [1] is

$$\dot{x} = a_{11}x(1 - [x/x_{max}]) + a_{12}x(1 - [y/y_{max}]), \tag{20.1}$$

$$\dot{y} = a_{21}y(1 - [y/y_{max}]) + a_{22}y(1 - [x/x_{max}]), \tag{20.2}$$

for initial conditions $x(0) = x_0$ and $y(0) = y_0$. In these nonlinear differential equations $\dot{x}$ ($\dot{y}$) is the time derivative of x (y). $x(t)$ ($y(t)$) is the population size of the first (second) species, and the maximum population size (carrying capacity) of x (y) is x_{max} (y_{max}). Without adding a carrying capacity both populations will grow without bound. The a_{ij} are all positive constants.

If we compare this model to the Predator/Prey Model in Chapter 7 we see that in the predator/prey equations some of the coefficients are negative and now for the symbiosis case all the coefficients are positive.

20.2 Parameters

There are eight parameters in equations (20.1) and (20.2) including the initial conditions. Let us assume that $x_{max} = y_{max} = 1000$ and $x_0 = y_0 = 5$ all known and crisp (not fuzzy). For a certain two species symbiosis under study the ecologists agree on values for $a_{11} \approx 0.3$ and $a_{21} \approx 0.4$ but they have been arguing about what to put down for a_{12} and a_{22}. Using expert opinion (Chapter 3) we arrive at triangular fuzzy numbers estimators for both a_{12} and a_{22}. All the values to be used for the parameters are now given in Table 20.1 including fuzzy values for a_{11} and a_{21}

Parameter	Fuzzy/Crisp Value
$\overline{a}_{11}$	(0.29/0.30/0.31)
$\overline{a}_{12}$	(0.15/0.17/0.19)
$\overline{a}_{21}$	(0.39/0.40/0.41)
$\overline{a}_{22}$	(0.18/0.20/0.22)
x_{max}	1000
y_{max}	1000
x_0	5
y_0	5

Table 20.1: Fuzzy/Crisp Parameter Values in the Symbiosis Example

Using fuzzy numbers for a_{ij} makes the system of differential equations into a system of fuzzy differential equations. The solutions will be fuzzy trajectories $\overline{x}(t)$ and $\overline{y}(t)$ so that any cut through the fuzzy trajectories for fixed t produces a fuzzy number. We want to find the band of maximum uncertainty which is the graph of the bases of the fuzzy numbers, or $\overline{x}(t)[0]$ and $\overline{y}(t)[0]$, the $\alpha = 0$ cut.

Next we need to choose the values of the fuzzy parameters in their $\alpha = 0$ cut to approximate the outer boundary of the $\alpha = 0$ cut of the fuzzy trajectory for $\overline{x}(t)$ and $\overline{y}(t)$. As discussed in Chapter 6 if $\overline{\omega} = (\omega_1/\omega_2/\omega_3)$ is a fuzzy parameter we use $\omega = \omega_1$, ω_2 and ω_3. We do this for $\overline{a}_{ij}$, $i, j = 1, 2$, giving $3^4 = 81$ curves for $x(t)$ and 81 graphs for $y(t)$ for a minimal approximation to $\overline{x}(t)[0]$ and $\overline{y}(t)[0]$. These are shown in Figures 20.2 and 20.3. The system diagram for Simulink is in Figure 20.1.

In Figure 20.2 (20.3) the reader can determine, from consulting Section 1.5.4, $x(t)$ ($y(t)$) when we used for the fuzzy parameters: (1) all left end points of their $\alpha = 0$ cut; (2) all right end points of their $\alpha = 0$ cut; and (3) their $\alpha = 1$ values.

20.3 Simulation

Let us first discuss the Simulink diagram in Figure 20.1. The input to "Integrate x' " is the right side of equation (20.1) and the input to "Integrate y'" is the right side of equation (20.2). The output from "Integrate x' " goes to: (1) the upper loop which computes $a_{11}x(1 - \frac{x}{1000})$; (2) the lower loop with gets $a_{12}x(1 - \frac{y}{1000})$; and (3) "Scope x" for its graph. Similar calculations are done for y in the lower half of the figure. "To X File" collects all the data on $x(t)$ so that we may present all the graphs in one figure. More details on this is in Chapter 28. There are also other items to set/choose, like type of numerical integrator and step size, which we will not discuss here (see Chapter 28).

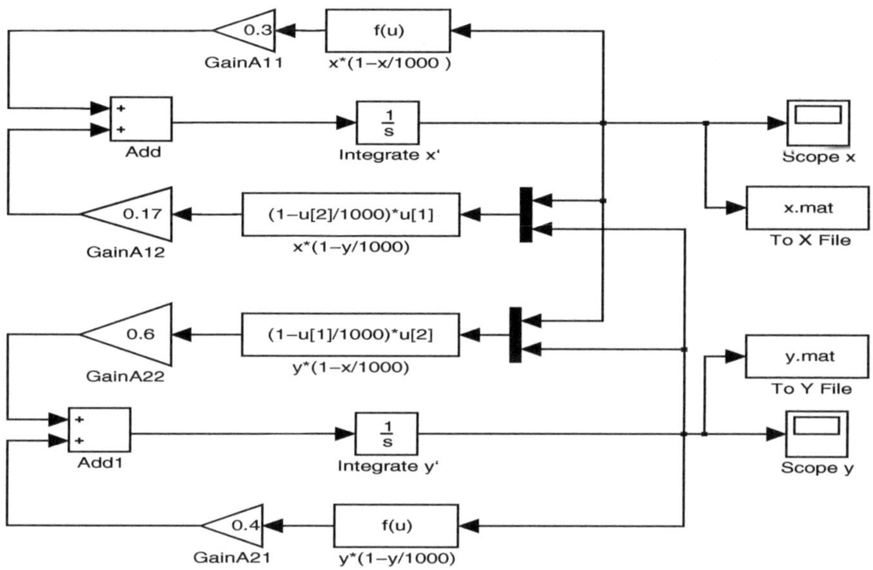

Figure 20.1: Simulink Diagram for the Symbiosis Example

Let $\overline{x}(t)[0] = [x_1(t,0), x_2(t,0)]$. In Figure 20.2 connect the upper boundary with a smooth curve which is our approximation to $x_2(t,0)$ and connect the lower boundary with a smooth curve producing an approximation to $x_1(t,0)$. The horizontal axis is time and the vertical axis is the number of species x. We get steady state (horizontal curves) with $\lim_{t\to\infty} x_i(t,0) = 1000$, $i = 1,2$. There is no uncertainty in this limit. Also, the x population gets to its maximum in about 45 time units.

Let $\overline{y}(t)[0] = [y_1(t,0), y_2(t,0)]$. In Figure 20.3 connect the upper boundary with a smooth curve which is our approximation to $y_2(t,0)$ and connect the lower boundary with a smooth curve producing an approximation to $y_1(t,0)$. The horizontal axis is time and the vertical axis is the number of species y. We get steady state (horizontal curves) with $\lim_{t\to\infty} y_i(t,0) = 1000$, $i = 1,2$. No uncertainty in the limit. The $y(t)$ population arrives at its carrying capacity in approximately 40 time units after over shooting it to to a maximum between 1100 to 1280.

If all the uncertain parameters are estimated from data (Chapter 3), without using expert opinion, then the base of the fuzzy estimator is a 99% confidence interval. Using all of these confidence intervals we get $\overline{x}(t)[0]$ and $\overline{y}(t)[0]$, so these bands of uncertainty will be like 99% confidence intervals for $x(t)$ and $y(t)$, respectively.

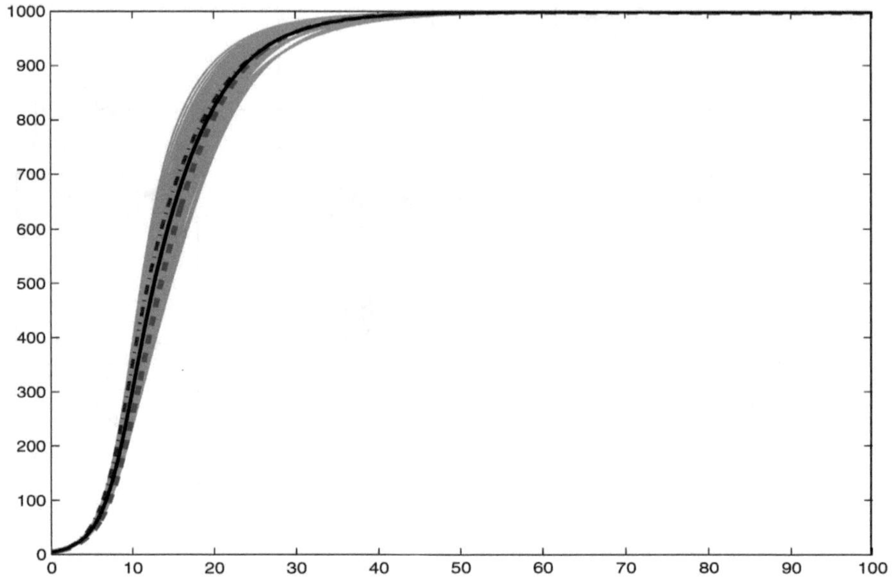

Figure 20.2: Fuzzy Trajectory for $\overline{x}(t)[0]$ in the Symbiosis Example

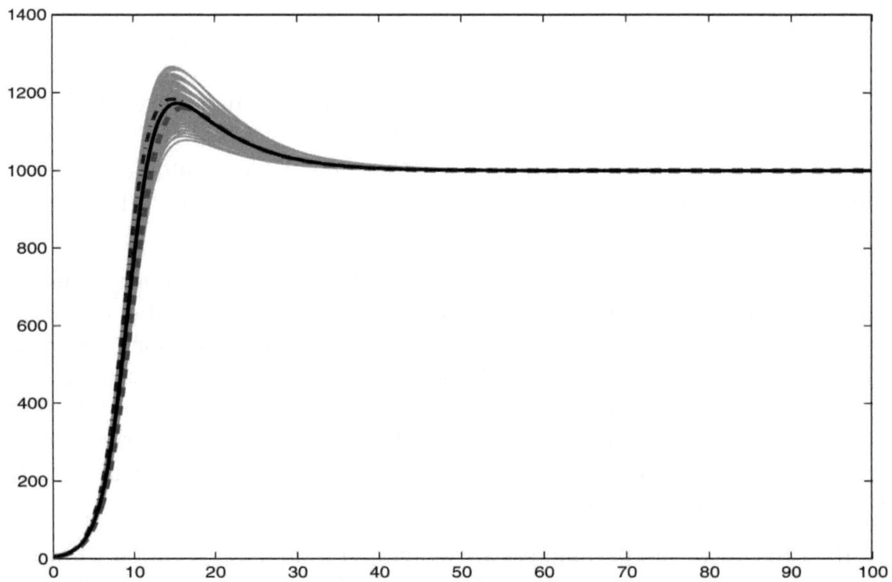

Figure 20.3: Fuzzy Trajectory for $\overline{y}(t)[0]$ in the Symbiosis Example

20.4 Reference

1. J.D.Murry: Mathematical Biology, Springer-Verlag, Heidelberg, Germany, 1989.

Chapter 21

Supply and Demand

21.1 Introduction

This simple mathematical model in economics is adapted from an example in [1]. The linear differential equations for this example are

$$\dot{P} = F(t) - k_1(S - S_0), \tag{21.1}$$

$$\dot{S} = k_2(P - P_0), \tag{21.2}$$

where $\dot{P}$ ($\dot{S}$) is the time derivative of $P = P(t) =$ price ($S = S(t) =$ supply), and initial conditions are $P(0) = P_0$, $S(0) = S_0$. It is assumed that the initial conditions (P_0, S_0) represent equilibrium price and supply.

Two items that will effect price are inflation $F(t)$ and supply S. Let us assume that inflation oscillates and we model inflation using a sine curve $Asin(Bt)$ in equation (21.1). If $S > S_0$ then supply is too large and price will decrease, but if $S < S_0$ then supply is too low and price will increase. So we add the factor $-k_1(S - S_0)$ to equation (21.1) with constant $k_1 > 0$. In equation (21.2) we have the term $k_2(P - P_0)$, for $k_2 > 0$, because we assume that: (1) if $P > P_0$ price is too high and supply S will increase; and (2) if $P < P_0$ price is too low and supply S will decrease.

These are linear differential equations so we could easily find the solutions. However, even using simple solutions it can be complicated to find the boundary of the band of uncertainty, or the boundary of $\overline{P}(t)[0]$) and $\overline{S}(t)[0]$, as shown in Section 6.2. So we omit expressions for the exact solutions and will use simulation instead.

21.2 Parameters

The parameters are k_1, k_2, P_0 , S_0, A and B. We assume that only inflation is uncertain so that k_1, k_2, P_0 and S_0 are all known and crisp but A and

Parameter	Crisp/Fuzzy Value
k_1	2
k_2	2
P_0	70
S_0	200
A	$\overline{A} = (10/20/30)$
B	$\overline{B} = (3/4/5)$

Table 21.1: Crisp/Fuzzy Parameters in the Supply and Demand Model

B are unknown and will be triangular fuzzy numbers. Using expert opinion (Chapter 3) we obtain fuzzy estimators for A and B. Fuzzy inflation does not seem too unreasonable. All the values of the constants are given in Table 21.1.

Using fuzzy parameters we get fuzzy differential equations and the solutions $\overline{P}(t)$ and $\overline{S}(t)$ are also fuzzy. The system now becomes a continuous fuzzy system whose trajectories are fuzzy so that any slice through a trajectory at some time t_z is a fuzzy number. We wish to estimate the band of uncertainty which is $\overline{P}(t)[0]$ and $\overline{S}(t)[0]$, or the $\alpha = 0$ cuts.

Next we need to choose the values of the fuzzy parameters in their $\alpha = 0$ cut to approximate the outer boundary of the $\alpha = 0$ cut of the fuzzy trajectory for price and supply. As discussed in Chapter 6 if $\overline{\omega} = (\omega_1/\omega_2/\omega_3)$ is a fuzzy parameter we use $\omega = \omega_1$, ω_2 and ω_3. We do this for $\overline{A}$ and $\overline{B}$ giving $3^2 = 9$ curves for $P(t)$ and 9 graphs for $S(t)$ for a minimal approximation to $\overline{P}(t)[0]$ and $\overline{S}(t)[0]$. These are shown in Figures 21.2 and 21.3. The fuzzy input for inflation is shown in Figure 21.4. The system diagram for Simulink is in Figure 21.1.

In Figures 21.2-21.4 the reader can determine, from consulting Section 1.5.4, $P(t)$, $S(t)$ and $F(t)$ (inflation) when we used for the fuzzy parameters: (1) all left end points of their $\alpha = 0$ cut; (2) all right end points of their $\alpha = 0$ cut; and (3) their $\alpha = 1$ values.

21.3 Simulation

Let us first discuss the Simulink diagram in Figure 21.1. The input to "Integrator" is the right side of equation (21.1) and the output is $P(t)$. $P(t)$ is sent to "Scope P" for its graph and also to "Subtract1". The input to "Subtract" is $Asin(Bt)$ and $k_1(S - S_0)$. The input to "Subtract1" is $P(t)$ and P_0 whose output gets multiplied by k_2 for input to "Integrate S' ". The "Scopes" are for single graphs and the "Files" collect all the data so we can place all graphs in one figure. More details on how this is done is in Chapter 28. There are also other items to set/choose, like simulation time, initial step size and type

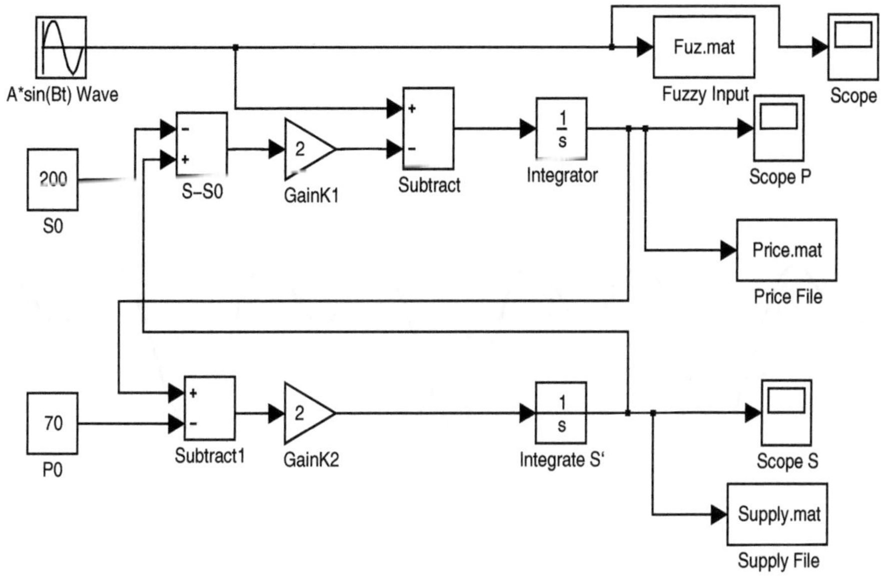

Figure 21.1: Simulink Diagram for the Supply and Demand Example

of numerical integrator, which we will not discuss here (see Chapter 28).

Let $\overline{P}(t)[0] = [p_1(t,0), p_2(t,0)]$. In Figure 21.2 connect the upper boundary with a smooth curve which is our approximation to $p_2(t,0)$ and connect the lower boundary with a smooth curve producing an approximation to $p_1(t,0)$. The horizontal axis is time and the vertical axis is \$. The fuzzy graph appears periodic. The possible values for price vary from about \$41 to approximately \$106 which is a large spread.

Let $\overline{S}(t)[0] = [s_1(t,0), s_2(t,0)]$. In Figure 21.3 connect the upper boundary with a smooth curve which is our approximation to $s_2(t,0)$ and connect the lower boundary with a smooth curve producing an approximation to $s_1(t,0)$. The horizontal axis is time and the vertical axis is the number of units supplied. The fuzzy graph looks periodic. Supply can range from a low of around 172 to a high of approximately 228. Also a large range of values.

Figure 21.4 shows fuzzy inflation modeled as a fuzzy sine wave. Fuzzy inflation causes the large variability in price and supply.

If all the uncertain parameters are estimated from data (Chapter 3), and not by employing expert advice, then the base of the fuzzy estimators is a 99% confidence interval. Using all of these confidence intervals we get $\overline{P}(t)[0]$ and $\overline{S}(t)[0]$, so these bands of uncertainty will be like 99% confidence intervals for $P(t)$ and $S(t)$, respectively.

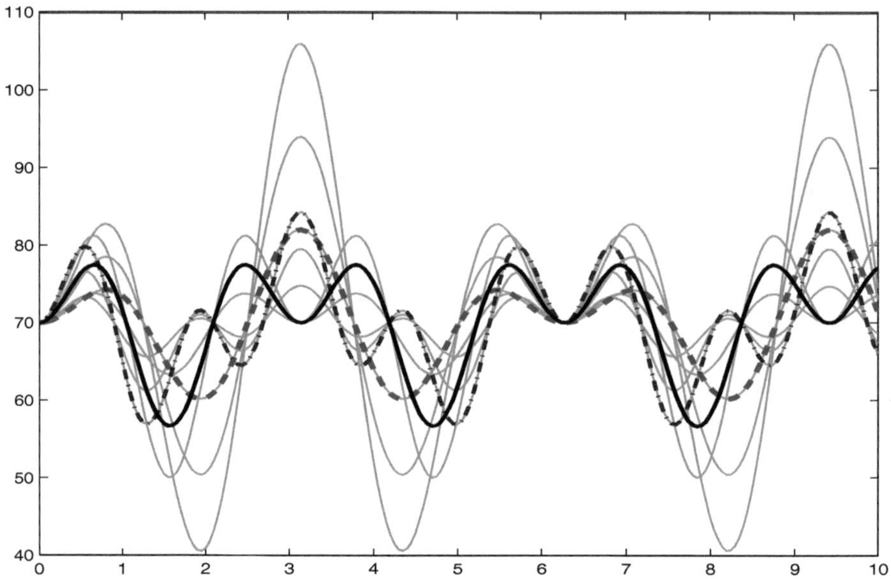

Figure 21.2: Fuzzy Trajectory for Price $\overline{P}(t)[0]$ in the Supply and Demand Example

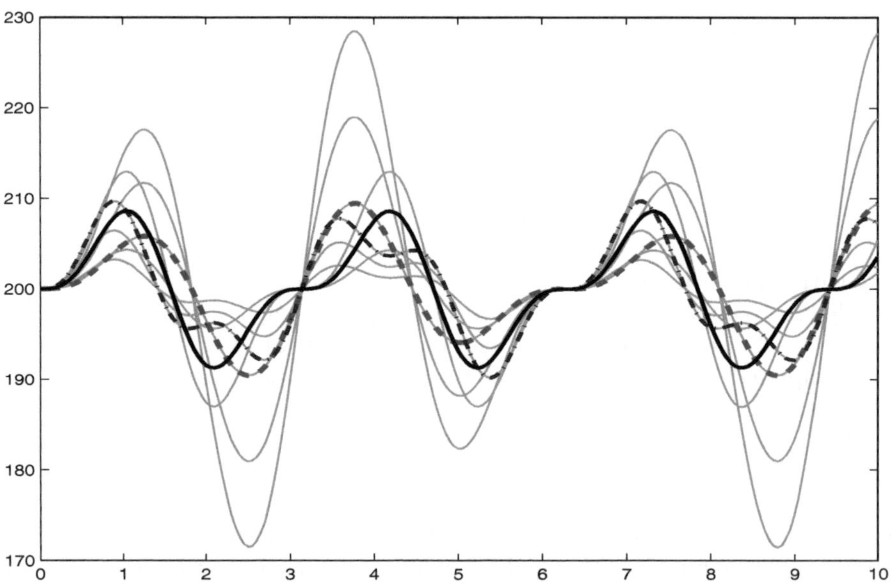

Figure 21.3: Fuzzy Trajectory for Supply $\overline{S}(t)[0]$ in the Supply and Demand Example

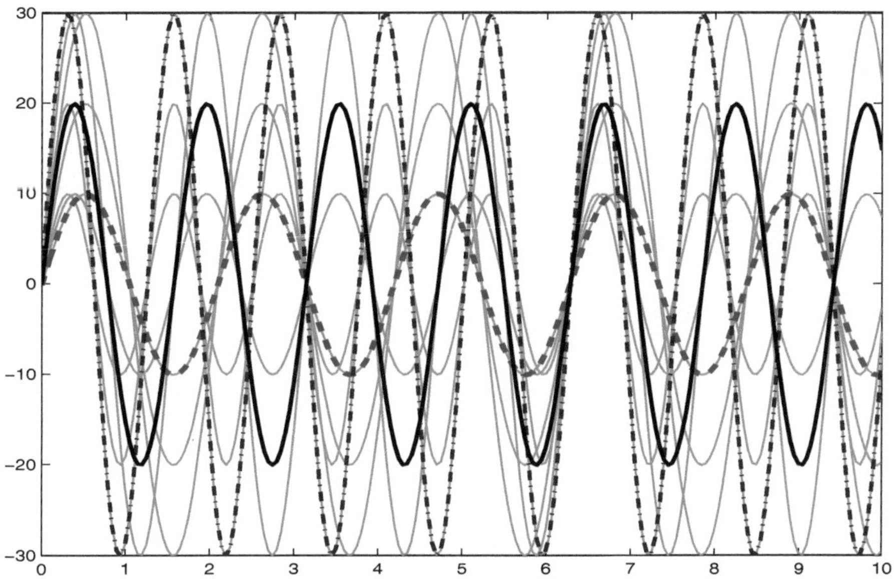

Figure 21.4: Fuzzy Graph for Inflation in the Supply and Demand Example

21.4 Reference

1. M.R.Spiegel: Applied Differential Equations, Third Edition, Prentice-Hall, Englewood Cliffs, NJ, 1981.

Chapter 22

Drug Concentrations

22.1 Introduction

This application has been adopted from an example in [1]. This will be a two compartment model, see Figure 22.1, but could be expanded to more compartments. Let V_i = the volume of compartment $i = 1, 2$ and let $x_i(t) =$ be the amount of the drug in compartment $i = 1, 2$ at time t. In Figure 22.1 A is the area of the membrane between the two compartments. The drug can flow through the membrane from compartment #1 to #2, but also from compartment #2 to #1. Figure 22.1 shows that the drug can also escape to the external system through an opening in compartment #2.

We have the rate of change of x_1 equals the rate of flow from #2 to #1 minus the rate of flow from #1 to #2. Let us assume that the rate of flow from #2 to #1 is proportional to A and the concentration (x_2/V_2) of the drug in compartment #2. Let this constant of proportionality be a_{21}. Then we use

$$a_{21}A(x_2/V_2), \tag{22.1}$$

for the rate of flow of the drug from #2 to #1. Similarly, we use $a_{12}A(x_1/V_1)$, for constant of proportionality a_{12}, to model the rate of flow from #1 to #2. Hence, our first differential equation is

$$\dot{x}_1 = a_{21}A(x_2/V_2) - a_{12}A(x_1/V_1). \tag{22.2}$$

The rate of change of x_2 equals the rate of flow from #1 into #2 minus the fate of flow from #2 to #1 and also minus the rate of flow out of #2 to the external system. Assume that the rate of flow from #2 to the external system is proportional to the concentration x_2/V_2 since there is no membrane there. Let this constant of proportionality be a_2. Then our second differential equation, obtained in a similar manner as the first equation, is

$$\dot{x}_2 = a_{12}A(x_1/V_1) - a_{21}A(x_2/V_2) - a_2(x_2/V_2). \tag{22.3}$$

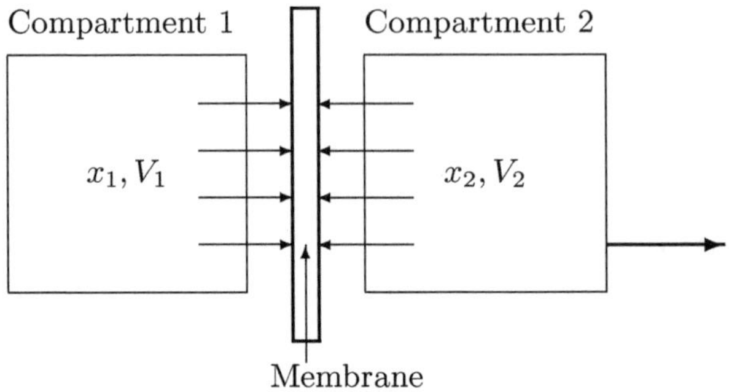

Figure 22.1: Diagram for the Drug Concentration Example

The system of differential equations is equations (22.2) and (22.3) together with initial conditions $x_1(0) = x_{10}$ and $x_2(0) = x_{20}$.

We will first combine some of the constants to simplify the equations. Let $b_{12} = a_{12}A/V_1$, $b_{21} = a_{21}A/V_2$ and $b_2 = a_2/V_2$. Then the system of linear differential equations becomes

$$\dot{x}_1 = b_{21}x_2 - b_{12}x_1, \tag{22.4}$$

$$\dot{x}_2 = b_{12}x_1 - (b_{21} + b_2)x_2. \tag{22.5}$$

These are linear differential equations so we could easily find the solutions. However, even using simple solutions it can be complicated to find the boundary of the band of uncertainty, or the boundary of $\overline{x}_1(t)[0]$ and $\overline{x}_2(t)[0]$, as shown in Section 6.2. So we omit expressions for the exact solutions and will use simulation instead.

22.2 Parameters

The parameters are b_{12}, b_{21}, b_2 , x_{10} and x_{20}. We assume that all the "rates" are uncertain so that x_{10} and x_{20} are known and crisp but b_{12}, b_{21} and b_2 are unknown and will be triangular fuzzy numbers. Actually, a_{12}, a_{21} and a_2 are the rates and are to be estimated from data as in Section 3.4, producing fuzzy number estimators. From these fuzzy numbers we obtain fuzzy numbers for the b_{12}, b_{21} and b_2 since we assume that values for A and the V_i are all known and crisp. For example, $\overline{b}_{12} = (A/V_1)\overline{a}_{12}$. All the values of the parameters are given in Table 22.1. We use triangular fuzzy number estimators for b_{12}, b_{21} and b_2.

Using fuzzy parameters we get fuzzy differential equations and the solutions $\overline{x}_1(t)$ and $\overline{x}_2(t)$ are also fuzzy. The solutions will be fuzzy trajectories so that any cut through the fuzzy trajectories for fixed t produces a fuzzy

Parameter	Crisp/Fuzzy Value
b_{12}	$\overline{b}_{12} = (0.2/0.3/0.4)$
b_{21}	$\overline{b}_{21} = (0.4/0.5/0.6)$
b_2	$\overline{b}_2 = (0.05/0.10/0.15)$
x_{10}	100
x_{20}	10

Table 22.1: Crisp/Fuzzy Parameters in the Drug Concentration Model

number. We want to find the band of maximum uncertainty which is the graph of the bases of the fuzzy numbers, or $\overline{x}_1(t)[0]$ and $\overline{x}_2(t)[0]$ the $\alpha = 0$ cut.

Next we need to choose the values of the fuzzy parameters in their $\alpha = 0$ cut to approximate the outer boundary of the $\alpha = 0$ cut of the fuzzy trajectory for $\overline{x}_i(t)$, $i = 1, 2$. As discussed in Chapter 6 if $\overline{\omega} = (\omega_1/\omega_2/\omega_3)$ is a fuzzy parameter we use $\omega = \omega_1$, ω_2 and ω_3. We do this for $\overline{b}_{12}$, $\overline{b}_{21}$ and $\overline{b}_2$ giving $3^3 = 27$ curves for $x_1(t)$ and 27 graphs for $x_2(t)$ for a minimal approximation to $\overline{x}_1(t)[0]$ and $\overline{x}_2(t)[0]$. These are shown in Figures 22.3 and 22.4. The systems diagram for Simulink is in Figure 22.2.

In Figure 22.3 (22.4) the reader can determine, from consulting Section 1.5.4, $x_1(t)$ ($x_2(t)$) when we used for the fuzzy parameters: (1) all left end points of their $\alpha = 0$ cut; (2) all right end points of their $\alpha = 0$ cut; and (3) their $\alpha = 1$ values.

22.3 Simulation

We first discuss the Simulink diagram in Figure 22.2. The input to "Integrate x_1' " is the right side of equation (22.4) with output $x_1(t)$. The input to "Subtract" is $b_{12}x_1$ and $b_{21}x_2$. The input to "Subtract1" will be $b_{12}x_1$ and $(b_{21} + b_2)x_2$. The input to "Integrate x_2' " is the right side of equation (22.5). The "Scopes" are for single graphs while the "Files" collect the data so that we may do all the graphs in one figure. More details on how we do this is in Chapter 28. There are also other items to set/choose, like choosing a fixed step, or variable step, numerical integration program, which we will not discuss here (see Chapter 28).

Let $\overline{x}_1(t)[0] = [x_{11}(t, 0), x_{12}(t, 0)]$. In Figure 22.3 connect the upper boundary with a smooth curve which is our approximation to $x_{12}(t, 0)$ and connect the lower boundary with a smooth curve producing an approximation to $x_{11}(t, 0)$. The horizontal axis is time and the vertical axis is the amount of the drug in compartment #1. We see that $x_1(t)$ decreases from its initial value and goes to zero between 75 and 450 time units. The time of 450 may be too long until this drug is out of Compartment 1.

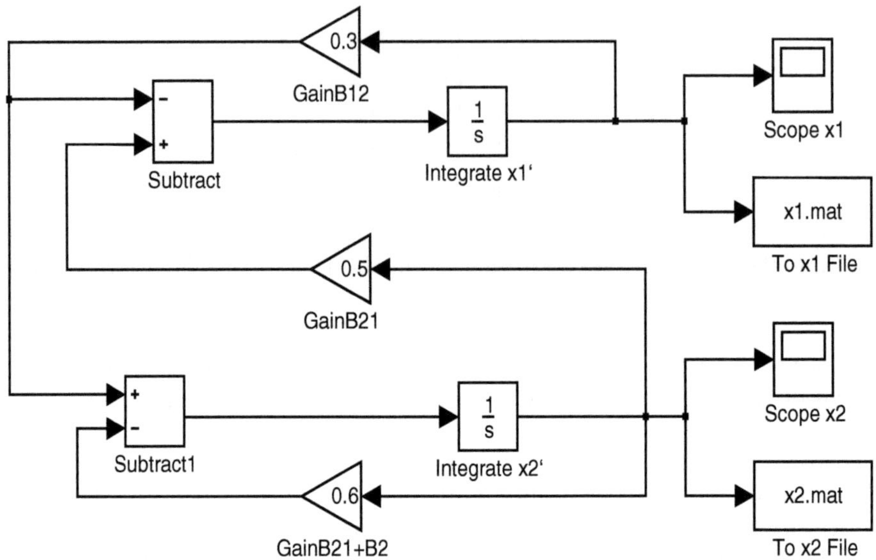

Figure 22.2: Simulink Diagram for the Drug Concentration Example

Let $\overline{x}_2(t)[0] = [x_{21}(t,0), x_{22}(t,0)]$. In Figure 22.4 connect the upper boundary with a smooth curve which is our approximation to $x_{22}(t,0)$ and connect the lower boundary with a smooth curve producing an approximation to $x_{21}(t,0)$. The horizontal axis is time and the vertical axis is the amount of the drug in compartment #2. $x_2(t)$ starts out at 10 units and rapidly increases to a maximum of between 22 and 47 units. Then $x_2(t)$ steadily decreases to zero and gets to zero between 80 and 400 time units. For this drug a maximum of 47 units may be too great and a time of 400 could take too long for it to leave Compartment 2.

If all the uncertain parameters are estimated from data (Chapter 3), and none come from expert opinion, then the base of the fuzzy estimator is a 99% confidence interval. Using all of these confidence intervals we get $\overline{x}_1(t)[0]$ and $\overline{x}_2(t)[0]$, so these bands of uncertainty will be like 99% confidence intervals for $x_1(t)$ and $x_2(t)$, respectively.

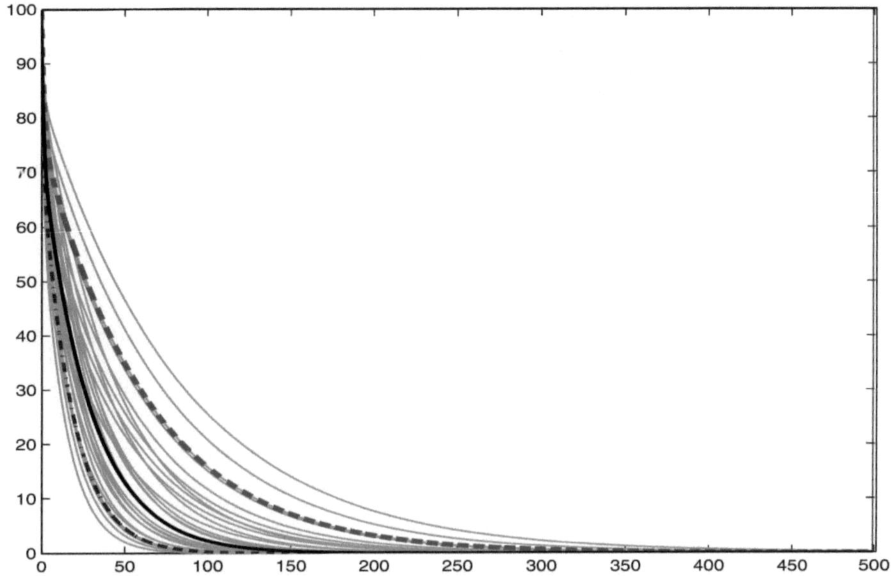

Figure 22.3: Fuzzy Trajectory, $\overline{x}_1(t)[0]$ in the Drug Concentration Example

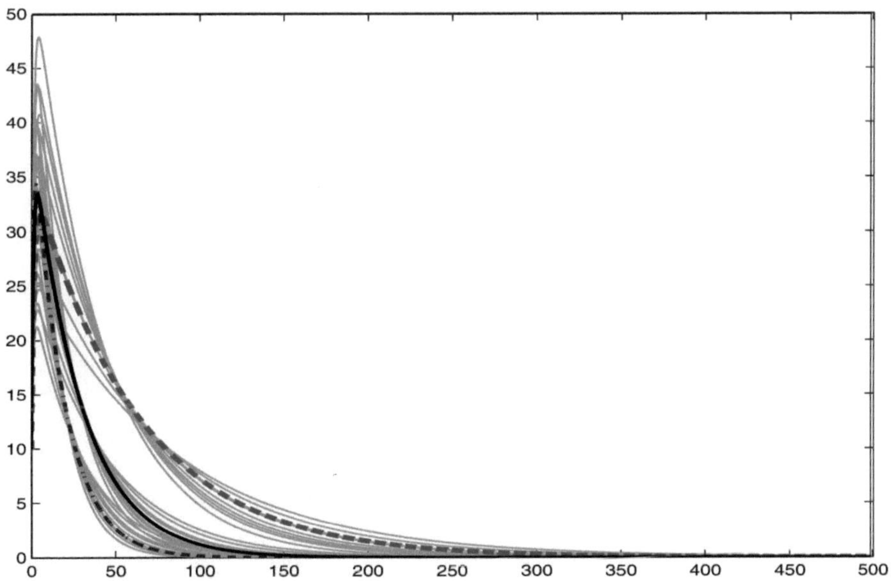

Figure 22.4: Fuzzy Trajectory, $\overline{x}_2(t)[0]$ in the Drug Concentration Example

22.4 Reference

1. M.R.Spiegel: Applied Differential Equations, Third Edition, Prentice-Hall, Englewood Cliffs, NJ, 1981.

Chapter 23

Three Species Competition

23.1 Introduction

This example generalizes the predator/prey model in Chapter 7 and has been adapted from ([1],[2]). Let $N_i(t)$ = the number of species i present at time $t \geq 0$, $i = 1, 2, 3$. Also let λ_i be the growth rate of species i in the absence of any competition, from any other species and from its own species, $i = 1, 2, 3$. Finally, define $a_{ij} > 0$ to be the effect of the presence of species j on the growth of species i, for $i, j = 1, 2, 3$. The system of nonlinear ordinary differential equations governing the evolution of this three species model is

$$\dot{N}_i = \lambda_i N_i (1 - \sum_{j=1}^{3} a_{ij} N_j), \tag{23.1}$$

for $i = 1, 2, 3$ with initial conditions $N_i(0) = N_{i0}$, $i = 1, 2, 3$. $\dot{N}_i$ is the time derivative of $N_i(t)$, $i = 1, 2, 3$.

There are fifteen parameters in the system and we shall make the following assumptions, as in [1], to reduce the number of constants to seven. Let: (1) $\lambda_1 = \lambda_2 = \lambda_3 = \lambda$; (2) $a_{12} = a_{23} = a_{31} = a > 0$; (3) $a_{21} = a_{32} = a_{13} = b > 0$; and (4) $a_{11} = a_{22} = a_{33} = c > 0$. Then the system of nonlinear differential equations becomes

$$\dot{N}_1 = \lambda N_1 (1 - cN_1 - aN_2 - bN_3), \tag{23.2}$$

$$\dot{N}_2 = \lambda N_2 (1 - bN_1 - cN_2 - aN_3), \tag{23.3}$$

$$\dot{N}_3 = \lambda N_3 (1 - aN_1 - bN_2 - cN_3). \tag{23.4}$$

23.2 Parameters

The parameters are λ, a, b, c, N_{10}, N_{20} and N_{30}. We assume that all the constants λ, a, b and c are uncertain but N_{10}, N_{20} and N_{30} are known and

Parameter	Crisp/Fuzzy Value
λ	$\overline{\lambda} = (0.14/0.15/0.16)$
a	$\overline{a} = (0.02/0.025/0.03)$
b	$\overline{b} = (0.02/0.025/0.03)$
c	$\overline{c} = (0.015/0.02/0.025)$
N_{10}	5
N_{20}	5
N_{30}	5

Table 23.1: Crisp/Fuzzy Parameters in the Three Species Competition Model

crisp. The rate λ is to be estimated from data, as in Section 3.4, producing a fuzzy number estimator. Expert opinion (Chapter 3) is used to estimate a, b and c. All the fuzzy estimators are to be triangular fuzzy numbers. The values of all the parameters are given in Table 23.1.

Using fuzzy parameters we get fuzzy differential equations and the solutions $\overline{N}_i(t)$, $i = 1, 2, 3$, are also fuzzy. The solutions will be fuzzy trajectories so that any cut through the fuzzy trajectories for fixed t produces a fuzzy number. We want to find the band of maximum uncertainty which is the graph of the bases of the fuzzy numbers, or $\overline{N}_i(t)[0]$, $i = 1, 2, 3$, the $\alpha = 0$ cut.

Next we need to choose the values of the fuzzy parameters in their $\alpha = 0$ cut to approximate the outer boundary of the $\alpha = 0$ cut of the fuzzy trajectory for $\overline{N}_i(t)$, $i = 1, 2, 3$. As discussed in Chapter 6 if $\overline{\omega} = (\omega_1/\omega_2/\omega_3)$ is a fuzzy parameter we use $\omega = \omega_1$, ω_2 and ω_3. We do this for $\overline{a}$, $\overline{b}$, $\overline{c}$ and $\overline{\lambda}$ giving $3^4 = 81$ curves for $N_i(t)$, $i = 1, 2, 3$, for a minimal approximation to $\overline{N}_i(t)[0]$, $i = 1, 2, 3$. These are shown in Figures 23.2 - 23.4. The systems diagram for Simulink is in Figure 23.1.

In Figures 23.2-23.4 the reader can determine, from consulting Section 1.5.4, $N_i(t)$, $i = 1, 2, 3$ when we used for the fuzzy parameters: (1) all left end points of their $\alpha = 0$ cut; (2) all right end points of their $\alpha = 0$ cut; and (3) their $\alpha = 1$ values.

23.3 Simulation

First we explain the Simulink diagram in Figure 23.1. The top loop computes N_1, the middle loop is for N_2 and the lowest loop gets N_3. The input to "Integrator" is the right side of equation (23.2). The input to "Dot Product" is the two vectors $(N_1(t), N_2(t), N_3(t))$ and (c, a, b) so the output will be $cN_1 + aN_2 + bN_3$ which, after subtracting from one, goes to "Product". The other two inputs to "Product" are λ and $N_1(t)$. So we now see how Simulink also obtains $N_2(t)$ and $N_3(t)$. The "Scopes" are for single graphs and the

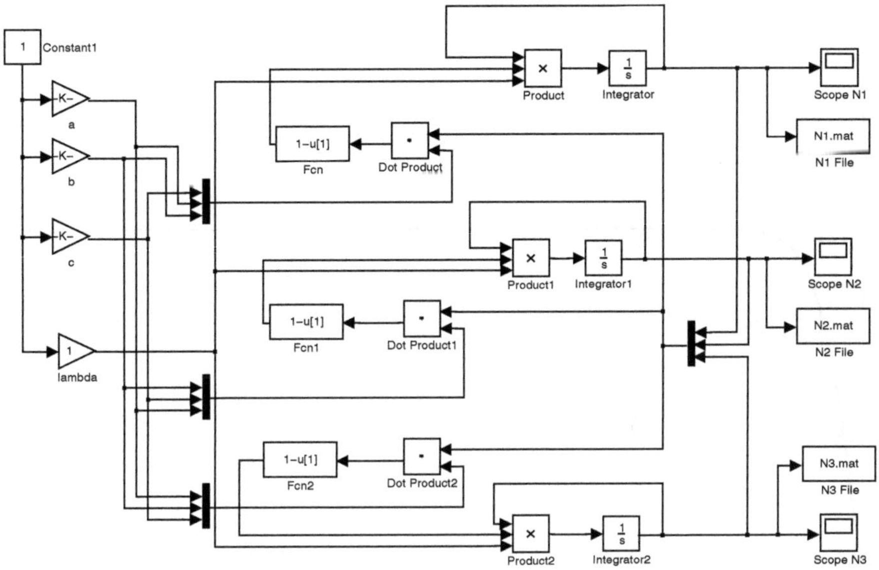

Figure 23.1: Simulink Diagram for the Three Species Competition

"Files" gather all the data so that we can put all graphs in one figure. Chapter 28 gives the code for this procedure. There are also other items to set/choose, like the type of numerical integrator and relative/absolute tolerance, which we will not discuss here (see Chapter 28).

Let $\overline{N}_1(t)[0] = [n_{11}(t,0), n_{12}(t,0)]$. In Figure 23.2 connect the upper boundary with a smooth curve which is our approximation to $n_{12}(t,0)$ and connect the lower boundary with a smooth curve producing an approximation to $n_{11}(t,0)$. The horizontal axis is time and the vertical axis is the number of species N_1. We get steady state (horizontal curves) with $\lim_{t\to\infty} n_{12}(t,0) \approx 18$ and $\lim_{t\to\infty} n_{11}(t,0) = 12$. The uncertainty band, the difference between the two curves in steady state, is approximately 6.

Let $\overline{N}_2(t)[0] = [n_{21}(t,0), n_{22}(t,0)]$. In Figure 23.3 connect the upper boundary with a smooth curve which is our approximation to $n_{22}(t,0)$ and connect the lower boundary with a smooth curve producing an approximation to $n_{21}(t,0)$. The horizontal axis is time and the vertical axis is the number of species N_2. We get steady state (horizontal curves) with $\lim_{t\to\infty} n_{22}(t,0) \approx 18$ and $\lim_{t\to\infty} n_{21}(t,0) = 12$. The uncertainty band, the difference between the two curves in steady state, is approximately 6.

The model is symmetric and we will obtain similar results for each species. Let $\overline{N}_3(t)[0] = [n_{31}(t,0), n_{32}(t,0)]$. In Figure 23.4 connect the upper boundary with a smooth curve which is our approximation to $n_{32}(t,0)$ and connect the lower boundary with a smooth curve producing an approx-

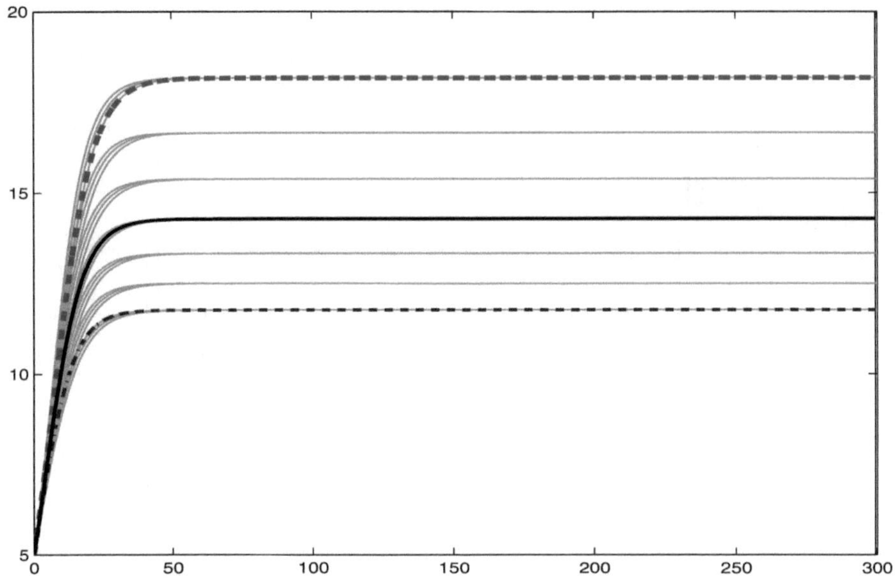

Figure 23.2: Fuzzy Trajectory for $\overline{N}_1(t)[0]$ in the Three Species Competition Example

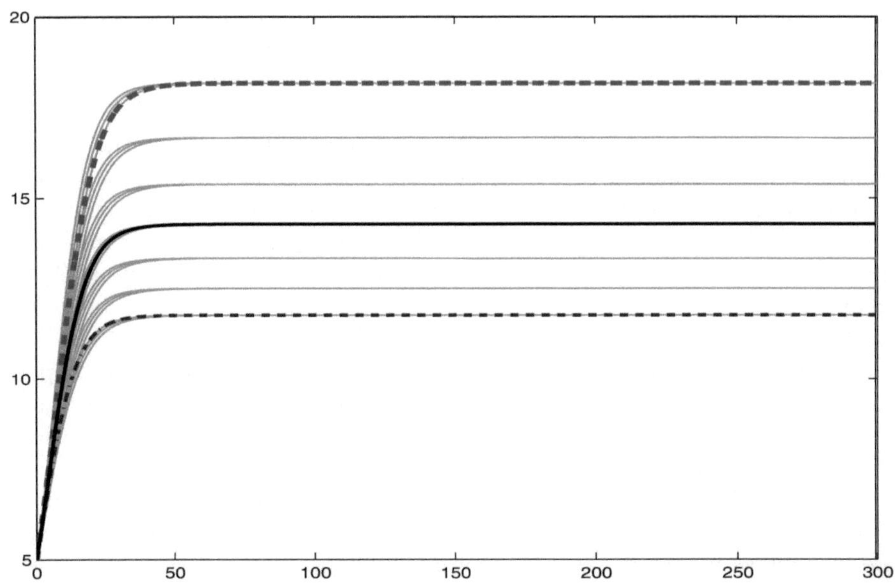

Figure 23.3: Fuzzy Trajectory for $\overline{N}_2(t)[0]$ in the Three Species Competition Example

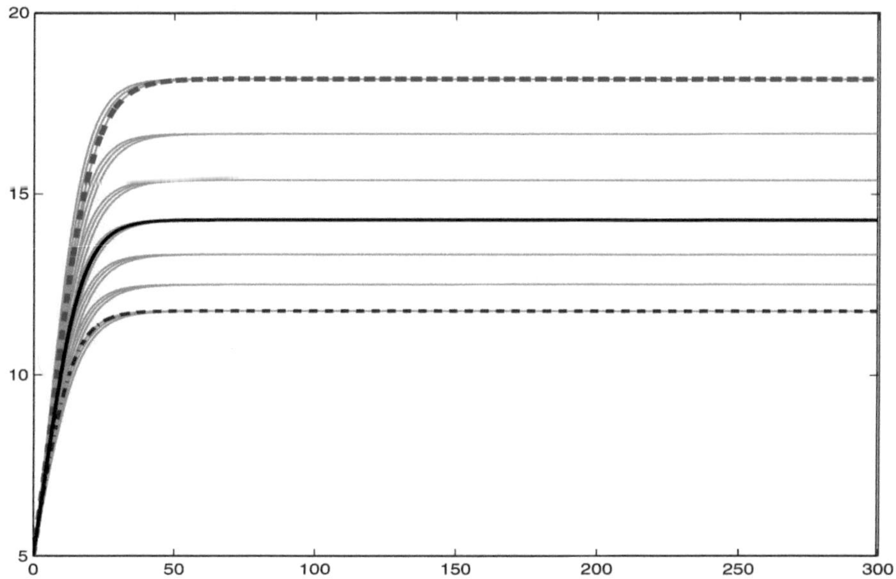

Figure 23.4: Fuzzy Trajectory for $\overline{N}_3(t)[0]$ in the Three Species Competition Example

imation to $n_{31}(t,0)$. The horizontal axis is time and the vertical axis is the number of species N_3. We get steady state (horizontal curves) with $\lim_{t\to\infty} n_{32}(t,0) \approx 18$ and $\lim_{t\to\infty} n_{31}(t,0) = 12$. The uncertainty band, the difference between the two curves in steady state, is approximately 6.

Although this model has four fuzzy parameter inputs the results show little uncertainty.

If all the uncertain parameters are estimated from data (Chapter 3), with none from expert advice, then the base of the fuzzy estimator is a 99% confidence interval. Using all of these confidence intervals we get $\overline{N}_i(t)[0]$, $i = 1, 2, 3$, so these bands of uncertainty will be like 99% confidence intervals for $N_i(t)$, $i = 1, 2, 3$.

23.4 References

1. J.C.Frauenthal: Introduction to Population Modeling, The UMAP Expository Monograph Series, Education Development Center, 1979.

2. R.M.May and W.J.Leonard: Nonlinear Aspects of Competition Between Three Species, SIAM J. Applied Mathematics, 29(1975)243-253.

Chapter 24

Flying a Glider

24.1 Introduction

This application was taken from a problem in ([1], p.188). Consider flying a glider at speed v making an angle θ with the horizontal. We measure θ positively in the counter clockwise direction and negatively in the clockwise direction. The system of nonlinear differential equations for the motion of the glider is approximated by

$$\dot{v} = -sin(\theta) - Dv^2, \tag{24.1}$$

$$v\dot{\theta} = -cos(\theta) + Dv^2, \tag{24.2}$$

where $\dot{v}$ ($\dot{\theta}$) is the time derivative of v (θ), $D > 0$ is the "drag" coefficient and the initial conditions are $v(0) = v_0$, $\theta(0) = \theta_0$

24.2 Parameters

The parameters are just D and the two initial conditions. Let us assume that the initial conditions are known and crisp. The drag coefficient D is difficult to obtain an exact value, so we ask some experts to estimate D (Chapter 3). The result is a triangular fuzzy number estimator for D. The values of the parameters are given in Table 24.1.

Using a fuzzy parameter $\overline{D}$ we get fuzzy differential equations and the solutions $\overline{v}(t)$ and $\overline{\theta}(t)$ are also fuzzy. The solutions will be fuzzy trajectories so that any cut through the fuzzy trajectories for fixed t produces a fuzzy number. We want to find the band of maximum uncertainty which is the graph of the bases of the fuzzy numbers, or $\overline{v}(t)[0]$ and $\overline{\theta}(t)[0]$ the $\alpha = 0$ cut.

Next we need to choose the values of the fuzzy parameters in their $\alpha = 0$ cut to approximate the outer boundary of the $\alpha = 0$ cut of the fuzzy trajectory for $\overline{v}(t)$ and $\overline{\theta}(t)$. As discussed in Chapter 6 if $\overline{\omega} = (\omega_1/\omega_2/\omega_3)$ is a

Parameter	Crisp/Fuzzy Value
D	$\overline{D} = (0.0005/0.0010/0.0015)$
v_0	132 ft/sec (90mph)
θ_0	-0.5236 rads (-30 degrees)

Table 24.1: Crisp/Fuzzy Parameters in the Glider Model

fuzzy parameter we (usually) use $\omega = \omega_1, \omega_2$ and ω_3. If we do this for $\overline{D}$ we get only 3 curves for $v(t)$ and 3 graphs for $\theta(t)$ for a very minimal approximation to $\overline{v}(t)[0]$ and $\overline{\theta}(t)[0]$. So let us choose more values for $D \in \overline{D}[0]$. We will use $D = 0.0005, 0.0006, ..., 0.0014, 0.0015$ for eleven values. Therefore we will get 11 curves for both $v(t)$ and $\theta(t)$. These are shown in Figures 24.2 and 24.3. The system diagram for Simulink is in Figure 24.1.

In Figure 24.2 (24.3) the reader can determine, from consulting Section 1.5.4, $v(t)$ $(\theta(t))$ when we used for D: (1)the left end point of its $\alpha = 0$ cut; (2) the right end point of its $\alpha = 0$ cut; and (3) its $\alpha = 1$ value.

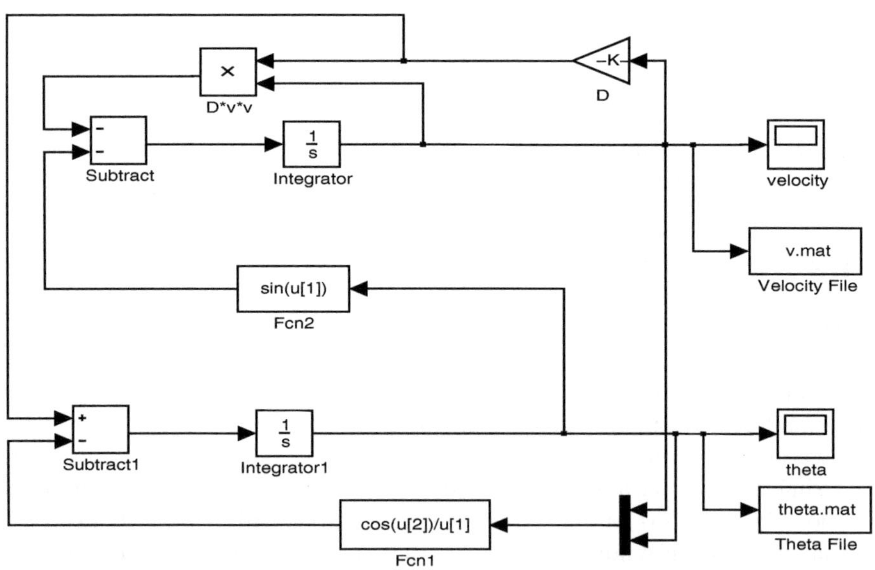

Figure 24.1: Simulink Diagram for the Glider Example

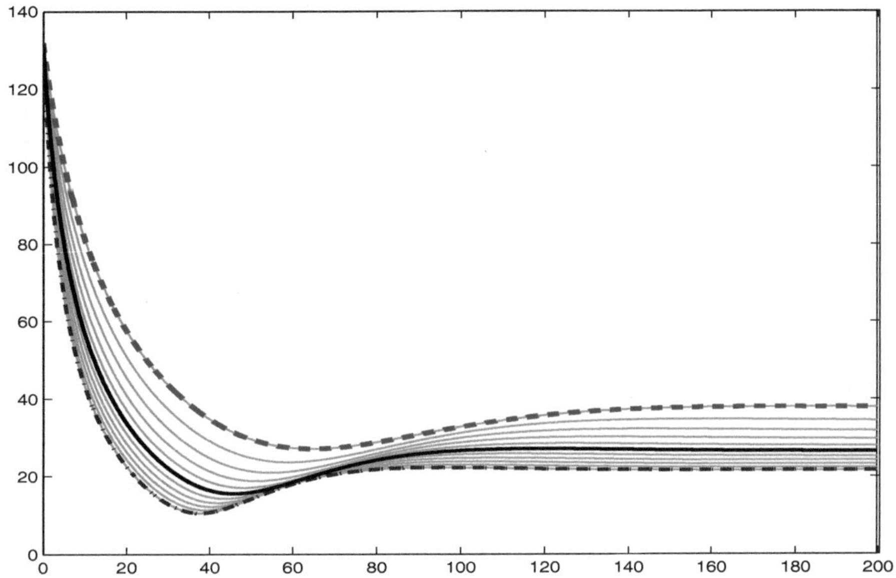

Figure 24.2: Fuzzy Trajectory for Velocity $\overline{v}(t)[0]$ in the Glider Example

24.3 Simulation

We first discuss how Simulink simulates the system using Figure 24.1. The top loop computes $v(t)$ and the bottom loop calculates $\theta(t)$. The input to "Integrator" is the right side of equation (24.1) and the input to "Integrator1" is the right side of equation (24.2) after solving for $\dot{\theta}$. The output from "Integrator" is $v(t)$ which travels to: (1) the top loop for calculating Dv^2 for input to "Subtract"; (2) "Fcn1" which computes $\frac{cos(\theta(t))}{v(t)}$ for input to "Subtract1" and "Integrator1"; (3) a "Scope" called "velocity" for its graph; and (4) "Subtract1". So we also see how $\theta(t)$ gets computed from equation (24.2). The boxes called "File" gathers all the data so that we may graph all eleven curves in one figure. Chapter 28 has these details. There are also other items to set/choose, like simulation time and type of numerical integrator, which we will not discuss here (see Chapter 28).

Let $\overline{v}(t)[0] = [v_1(t,0), v_2(t,0)]$. In Figure 24.2 connect the upper boundary with a smooth curve which is our approximation to $v_2(t,0)$ and connect the lower boundary with a smooth curve producing an approximation to $v_1(t,0)$. The horizontal axis is time (seconds) and the vertical axis is the velocity (feet/sec) of the glider. Assuming the glider is coming in for a landing we should plan on it between 40 and 80 seconds. The uncertainty band is not large and we could probably land anytime after 40 seconds.

Let $\overline{\theta}(t)[0] = [\theta_1(t,0), \theta_2(t,0)]$. In Figure 24.3 connect the upper boundary

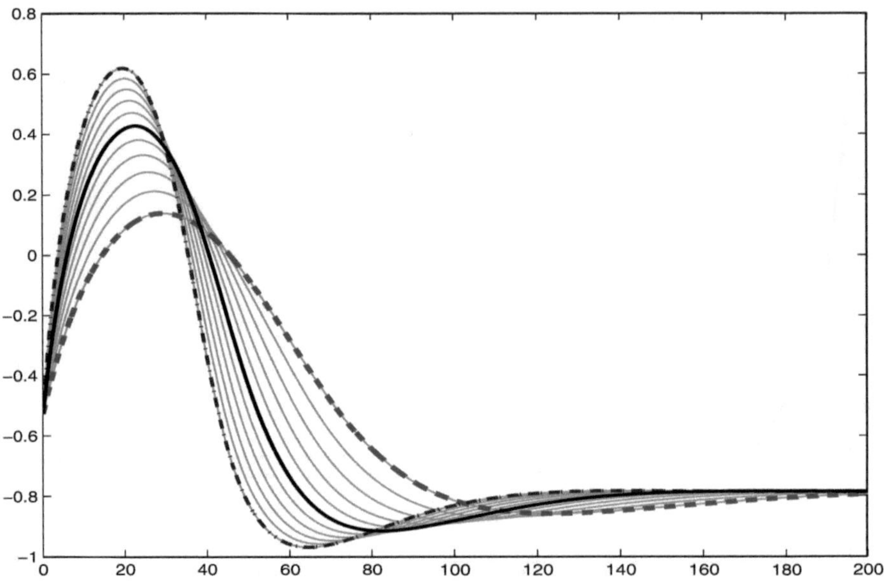

Figure 24.3: Fuzzy Trajectory for Angle $\overline{\theta}(t)[0]$ in the Glider Example

with a smooth curve which is our approximation to $\theta_2(t,0)$ and connect the lower boundary with a smooth curve producing an approximation to $\theta_1(t,0)$. The horizontal axis is time (seconds) and the vertical axis is the angle (in radians) the glider makes with the horizontal. We want to land when the nose of the glider is almost horizontal ($\theta \approx 0$). The uncertainty band shows that we can expect $\theta \approx 0$ for $t \approx 40$ seconds. Plan to land at 40 seconds. But, at about $t = 20$ seconds we have uncertainty $\theta = 0.10$ rads (+5.7 degrees) to $\theta = 0.62$ rads (+35 degrees). The nose of the glider up 35 degrees from the horizontal could be dangerous. But after 40 seconds we see it is possible for $\theta \approx -0.95$ (−54 degrees) which is certainly dangerous.

If all the uncertain parameters are estimated from data (Chapter 3), with none from expert advice, then the base of the fuzzy estimator is a 99% confidence interval. Using all of these confidence intervals we get $\overline{v}(t)[0]$ and $\overline{\theta}(t)[0]$, so these bands of uncertainty will be like 99% confidence intervals for $v(t)$ and $\theta(t)$.

24.4 Reference

1. S.H.Strogatz: Nonlinear Dynamics and Chaos, Addison-Wesley, Reading, MA., 1994.

Chapter 25

The National Economy

25.1 Introduction

This application was adopted from a problem in ([1], p.86). This is a simple (beginning) model for the national economy of some country. The system of linear differential equations is

$$\dot{I} = I - aC, \tag{25.1}$$

$$\dot{C} = b(I - C - G), \tag{25.2}$$

where I= the national income, C= the rate of consumer spending, G= the rate of government expenditures and $\dot{I}$ ($\dot{C}$) is the time derivative of I (C). Two of the parameters are a and b so that $a > 1$ and $b \geq 1$. We wish to investigate three cases: (1) case #1 $G = G_0$ a constant; (2) case #2 $G = G_0 + kI$ for $k > 0$; and (3) case #3 $G = G_0 + kI^{1.1}$ with $k > 0$. The initial conditions are $I(0) = I_0$ and $C(0) = C_0$.

These are linear differential equations so we could easily find the solutions. However, even using simple solutions it can be complicated to find the boundary of the band of uncertainty, or the boundary of $\overline{I}(t)[0]$) and $\overline{C}(t)[0]$, as shown in Section 6.2. So we omit expressions for the exact solutions and will use simulation instead.

25.2 Parameters

All the parameters are a, b, G_0, k and the two initial conditions. Let us assume that the government effect is fuzzy (a reasonable assumption) and the other parameters are all known and crisp. This means that a, b, I_0 and C_0 are all crisp but G_0 and k will be fuzzy. Our fuzzy estimators for G_0 and k will be triangular fuzzy numbers, obtained from expert opinion or from data (Chapter 3), and all the parameter values are given in Table 25.1.

Parameter	Crisp/Fuzzy Value
G_0	$\overline{G}_0 = (20,000/30,000/40,000)$
k	$\overline{k} = (0.04/0.05/0.06)$
a	1.5
b	1.0
I_0	$\$100,000$
C_0	$\$80,000$

Table 25.1: Crisp/Fuzzy Parameters in the National Economy Model

Using fuzzy parameters we get fuzzy differential equations and the solutions $\overline{I}(t)$ and $\overline{C}(t)$ are also fuzzy. The system now becomes a continuous fuzzy system whose trajectories are fuzzy so that any slice through a trajectory at some time t_z is a fuzzy number. We wish to estimate the band of uncertainty which is $\overline{I}(t)[0]$ and $\overline{C}(t)[0]$, or the $\alpha = 0$ cuts.

Next we need to choose the values of the fuzzy parameters in their $\alpha = 0$ cut to approximate the outer boundary of the $\alpha = 0$ cut of the fuzzy trajectory for the national income and the rate of consumer spending. As discussed in Chapter 6 if $\overline{\omega} = (\omega_1/\omega_2/\omega_3)$ is a fuzzy parameter we use $\omega = \omega_1$, ω_2 and ω_3. We do this for $\overline{G}_0$ and $\overline{k}$ giving three curves for case #1 and $3^2 = 9$ curves for $I(t)$ and $C(t)$ in the other cases, for a minimal approximation to $\overline{I}(t)[0]$ and $\overline{C}(t)[0]$. These are shown in Figures 25.2 - 25.7. The systems diagram for Simulink is in Figure 25.1.

In Figures 25.2-25.7 the reader can determine, from consulting Section 1.5.4, $I(t)$ ($C(t)$) when we used for the fuzzy parameters: (1) all left end points of their $\alpha = 0$ cut; (2) all right end points of their $\alpha = 0$ cut; and (3) their $\alpha = 1$ values.

25.3 First Simulation: Case #1

Here we assume that $G = \overline{G}_0$. We first discuss the Simulink diagram in Figure 25.1. The upper loop computes $I(t)$ and the lower loop gets $C(t)$. The input to "Integrator" is the right side of equation (25.1). The input to "Integrator1" is the right side of equation (25.2). Let us look at the bottom loop. The box "Add" calculates $kI + G_0 = G$ for input to "Subtract2". We have case #1 for $k = 0$. The other inputs to "Subtract2" are $I(t)$ and $C(t)$. The output from "Subtract2" gets multiplied by b before input to "Integrator1". The "Scopes" labeled "National Income" and "Consumer Spending" are for single graphs of I and C. The boxes labeled "File" collect all the data so that we can present all graphs in one figure. Chapter 28 has details on how this is accomplished. There are also other items to set/choose, like step size and type of numerical integrator, which we will not discuss here (see Chapter 28).

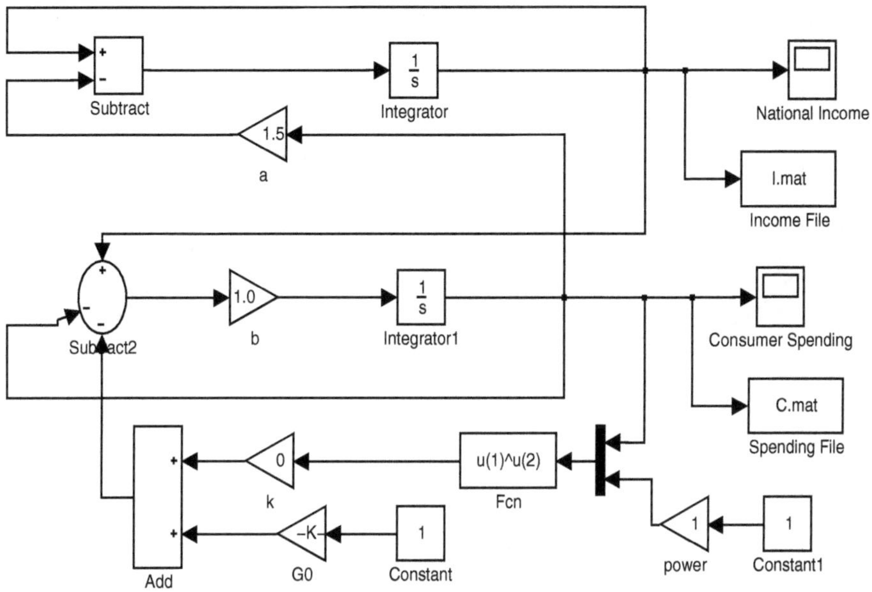

Figure 25.1: Simulink Diagram for the National Economy Model

Let $\overline{I}(t)[0] = [I_1(t,0), I_2(t,0)]$. In Figure 25.2 connect the upper boundary with a smooth curve which is our approximation to $I_2(t,0)$ and connect the lower boundary with a smooth curve producing an approximation to $I_1(t,0)$. The horizontal axis is time and the vertical axis is national income. We have oscillatory motion With $I(t)$ possibly varying between $10,000$ and $15,000$.

Let $\overline{C}(t)[0] = [c_1(t,0), c_2(t,0)]$. In Figure 25.3 connect the upper boundary with a smooth curve which is our approximation to $c_2(t,0)$ and connect the lower boundary with a smooth curve producing an approximation to $c_1(t,0)$. The horizontal axis is time and the vertical axis is the rate of consumer spending. We have oscillatory motion With $C(t)$ possibly varying between zero and $11,000$.

If all the uncertain parameters are estimated from data (Chapter 3), and none from expert opinion, then the base of the fuzzy estimator is a 99% confidence interval. Using all of these confidence intervals we get $\overline{I}(t)[0]$ and $\overline{C}(t)[0]$, so these bands of uncertainty will be like 99% confidence intervals for $I(t)$ and $C(t)$, respectively.

25.4 Second Simulation: Case #2

Now we assume that $G = \overline{G}_0 + \overline{k}I$.

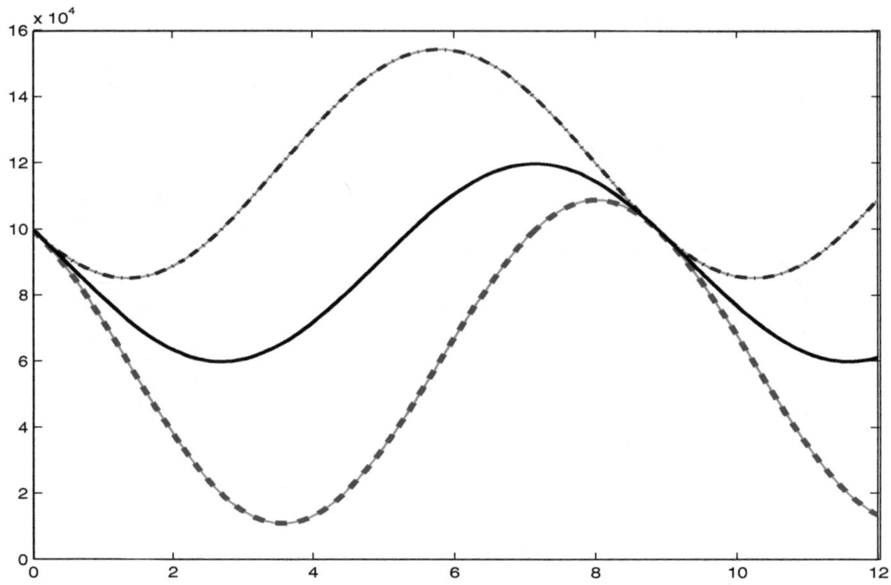

Figure 25.2: Fuzzy Trajectory for National Income $\overline{I}(t)[0]$ in the First National Economy Model

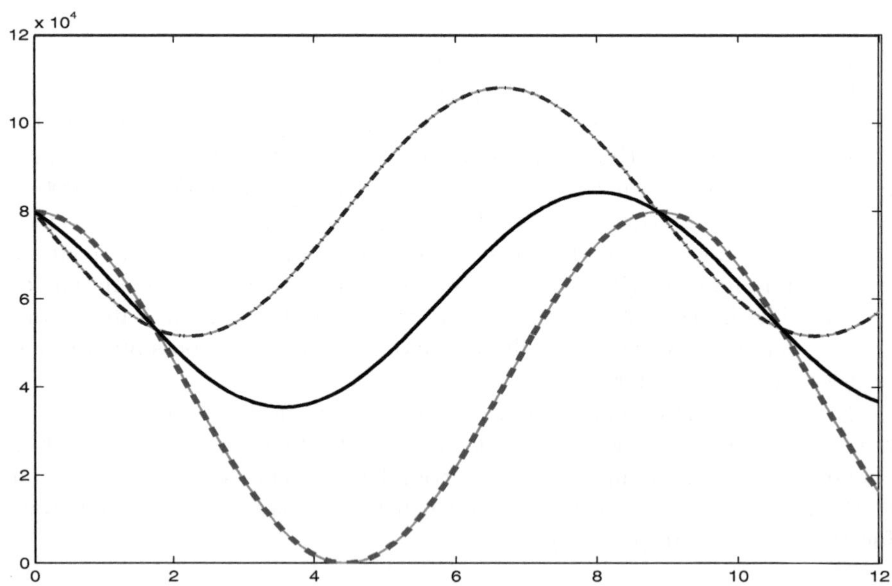

Figure 25.3: Fuzzy Trajectory for Consumer Spending $\overline{C}(t)[0]$ in the First National Economy Model

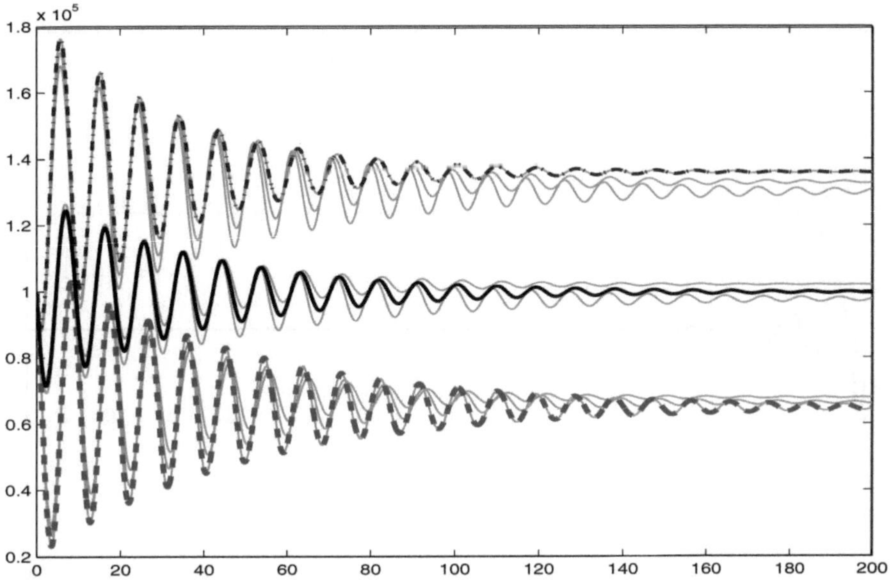

Figure 25.4: Fuzzy Trajectory for National Income $\overline{I}(t)[0]$ in the Second National Economy Model

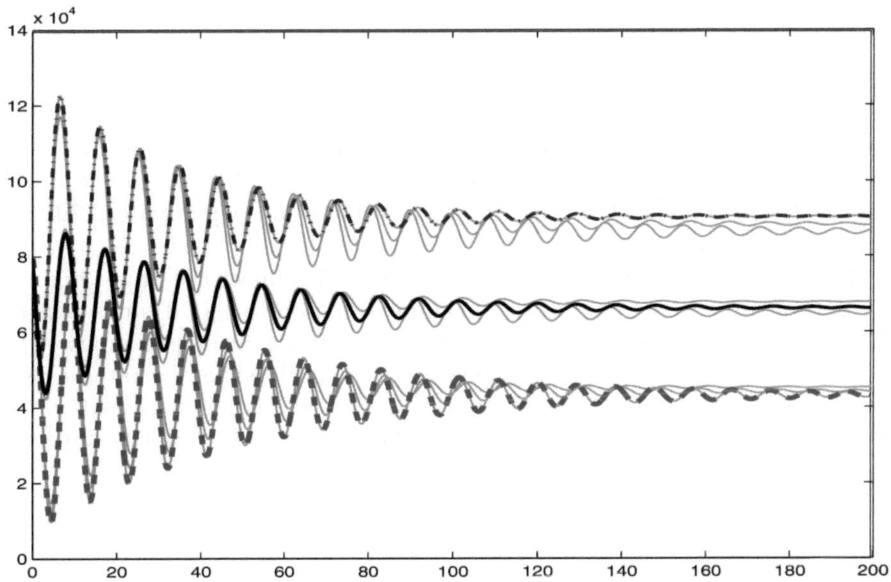

Figure 25.5: Fuzzy Trajectory for Consumer Spending $\overline{C}(t)[0]$ in the Second National Economy Model

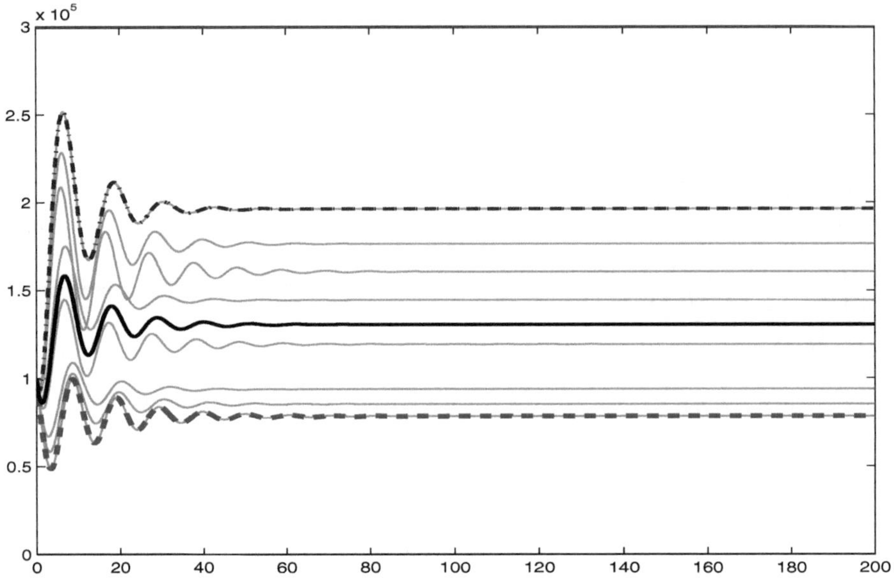

Figure 25.6: Fuzzy Trajectory for National Income $\overline{I}(t)[0]$ in the Third National Economy Model

Let $\overline{I}(t)[0] = [I_1(t,0), I_2(t,0)]$. In Figure 25.4 connect the upper boundary with a smooth curve which is our approximation to $I_2(t,0)$ and connect the lower boundary with a smooth curve producing an approximation to $I_1(t,0)$. The horizontal axis is time and the vertical axis is national income. We get a very interesting result. We see that $\lim_{t\to\infty} I_2(t,0) \approx 135,000$ and $\lim_{t\to\infty} I_1(t,0) \approx 65,000$. The uncertainty band is approximately $70,000$ which is quite large.

Let $\overline{C}(t)[0] = [c_1(t,0), c_2(t,0)]$. In Figure 25.5 connect the upper boundary with a smooth curve which is our approximation to $c_2(t,0)$ and connect the lower boundary with a smooth curve producing an approximation to $c_1(t,0)$. The horizontal axis is time and the vertical axis is the rate of consumer spending. Another interesting result. We see that $\lim_{t\to\infty} C_2(t,0) \approx 90,000$ and $\lim_{t\to\infty} C_1(t,0) \approx 45,000$. The uncertainty band is approximately $45,000$ which is somewhat large.

There is too much uncertainty and we should try to get less fuzzy estimators for G_0 and k.

25.5 Third Simulation: Case #3

Finally, we assume that $G = \overline{G}_0 + \overline{k}I^{1.1}$.

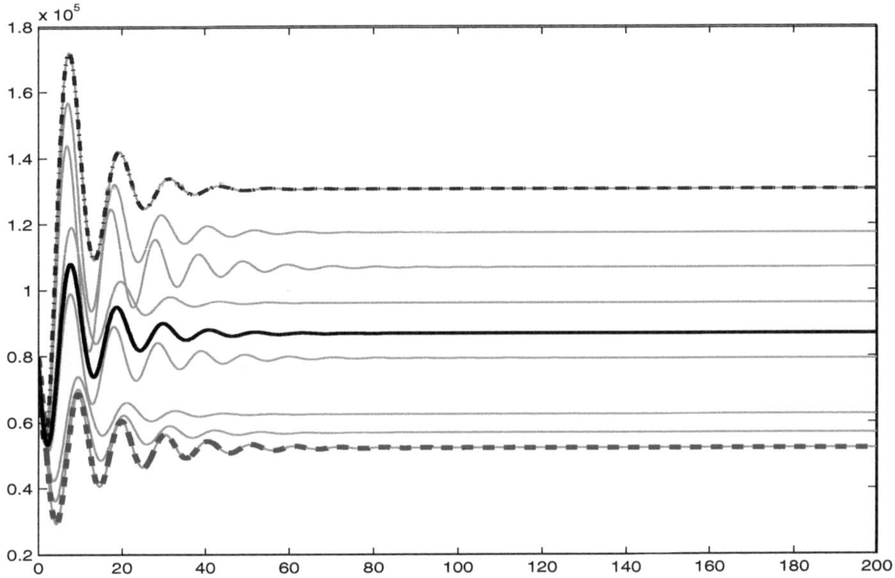

Figure 25.7: Fuzzy Trajectory for Consumer Spending $\overline{C}(t)[0]$ in the Third National Economy Model

Let $\overline{I}(t)[0] = [I_1(t,0), I_2(t,0)]$. In Figure 25.6 connect the upper boundary with a smooth curve which is our approximation to $I_2(t,0)$ and connect the lower boundary with a smooth curve producing an approximation to $I_1(t,0)$. The horizontal axis is time and the vertical axis is national income. We get an interesting result. We see that $\lim_{t\to\infty} I_2(t,0) \approx 200,000$ and $\lim_{t\to\infty} I_1(t,0) \approx 75,000$. The uncertainty band is approximately $125,000$ which is very large.

Let $\overline{C}(t)[0] = [c_1(t,0), c_2(t,0)]$. In Figure 25.7 connect the upper boundary with a smooth curve which is our approximation to $c_2(t,0)$ and connect the lower boundary with a smooth curve producing an approximation to $c_1(t,0)$. The horizontal axis is time and the vertical axis is the rate of consumer spending. Another interesting result. We see that $\lim_{t\to\infty} C_2(t,0) \approx 130,000$ and $\lim_{t\to\infty} C_1(t,0) \approx 53,000$. The uncertainty band is approximately $77,000$ which is quite large.

There is too much uncertainty and we should try to get less fuzzy estimators for G_0 and k.

25.6 Reference

1. D.W.Jordan and P.Smith: Nonlinear Ordinary Differential Equations, Third Edition, Oxford University Press, Oxford, UK, 2004.

Chapter 26

Sex Structured Population Models

26.1 Introduction

This model comes from Chapter 25 in [1]. The example we will look at is considered a "two-sex" model (we could look at three groups like unmarried females, unmarried males and couples) and an "age independent" model. The system of nonlinear differential equations to start with is

$$\dot{F} = -\mu_f F + b_f \Psi(F, M), \tag{26.1}$$

$$\dot{M} = -\mu_m M + b_m \Psi(F, M), \tag{26.2}$$

where $F(t)$ = the number of females in the population, $M(t)$ = the number of males in the population, $\mu_f F$ the number of females dying per time unit, $\mu_m M$ the number of males dying per time unit, $b_f \Psi(F, M)$ the number of female births per time unit, $b_m \Psi(F, M)$ the number of male births per time unit and $\Psi(F, M)$ some generally unknown function of F and M. The "dot" notation is for the time derivative. The population need not be people because we could be modeling squirrels, spiders, etc. The initial conditions are $F(0) = F_0$ and $M(0) = M_0$.

There are various choices for Ψ [1]: (1) $\Psi(F, M) = FM$ (law of mass action); (2) $\Psi(F, M) = F$ (female dominance); (3) $\Psi(F, M) = M$ (male dominance); (4) $\Psi(F, M) = \sqrt{FM}$ (geometric mean); (5) $\psi(F, M) = min(F, M)$ (min rule); and (6) $\psi(F, M) = \frac{2FM}{F+M}$ (harmonic mean). We will employ the harmonic mean function. So the system becomes

$$\dot{F} = -\mu_f F + b_f \frac{2FM}{F + M}, \tag{26.3}$$

$$\dot{M} = -\mu_m M + b_m \frac{2FM}{F + M}. \tag{26.4}$$

Parameter	Fuzzy/Crisp Value
b_f	$\overline{b}_f = (0.0074/0.0084/0.0094)$
b_m	$\overline{b}_m = (0.00769/0.00869/0.00969)$
μ_f	$\overline{\mu}_f = (0.0071/0.0081/0.0091)$
μ_m	$\overline{\mu}_m = (0.0080/0.0090/0.0100)$
F_0	$131,983,000$
M_0	$125,800,000$

Table 26.1: Fuzzy/Crisp Parameters in Sex Structured Population Model

26.2 Parameters

Let us now concentrate on modeling the population of people in some country. The parameters are μ_f, μ_m, b_f, b_m and the initial conditions. There are six parameters and we will consider four of them to be fuzzy. Assume that the initial conditions are all known and crisp so that μ_f, μ_m, b_f and b_m are not known precisely. Using expert opinion (Chapter 3) we get triangular fuzzy number estimators for these four parameters. The values of all the parameters are in Table 26.1.

Using fuzzy parameters we get fuzzy differential equations and the solutions $\overline{F}(t)$ and $\overline{M}(t)$ are also fuzzy. The system now becomes a continuous fuzzy system whose trajectories are fuzzy so that any slice through a trajectory at some time t_z is a fuzzy number. We wish to estimate the band of uncertainty which is $\overline{F}(t)[0]$ and $\overline{M}(t)[0]$, or the $\alpha = 0$ cuts.

Next we need to choose the values of the fuzzy parameters in their $\alpha = 0$ cut to approximate the outer boundary of the $\alpha = 0$ cut of the fuzzy trajectory for the population of females and males. As discussed in Chapter 6 if $\overline{\omega} = (\omega_1/\omega_2/\omega_3)$ is a fuzzy parameter we use $\omega = \omega_1$, ω_2 and ω_3. We do this for $\overline{\mu}_f$, $\overline{\mu}_m$, $\overline{b}_f$ and $\overline{b}_m$ giving $3^4 = 81$ curves for $F(t)$ and 81 graphs for $M(t)$ for a minimal approximation to $\overline{F}(t)[0]$ and $\overline{M}(t)[0]$. These are shown in Figures 26.2 and 26.3. The systems diagram for Simulink is in Figure 26.1.

In Figure 26.2 (26.3) the reader can determine, from consulting Section 1.5.4, $F(t)$ ($M(t)$) when we used for the fuzzy parameters: (1) all left end points of their $\alpha = 0$ cut; (2) all right end points of their $\alpha = 0$ cut; and (3) their $\alpha = 1$ values. However, in this application all three curves appear to coincide.

26.3 Simulation

We first discuss the Simulink diagram in Figure 26.1. The top loop computes $F(t)$ and the bottom loop is for $M(t)$. Both loops are similar so let us only talk about the top loop. The input to "Integrate F' " is the right

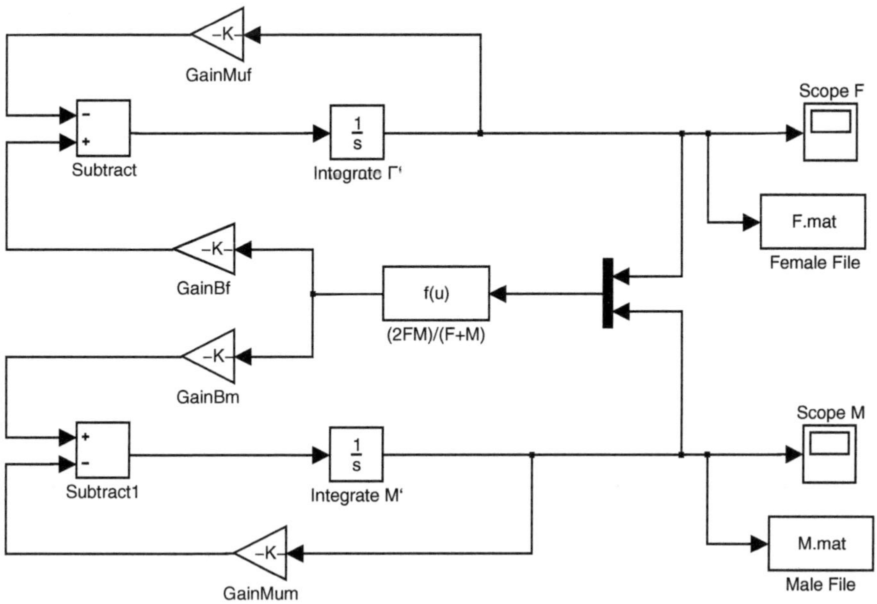

Figure 26.1: Simulink Diagram for the Sex Structured Population Model

side of equation (26.3) and its output is $F(t)$. $F(t)$ travels: (1) around a top loop where it gets multiplied by μ_f before input to "Subtract"; (2) to the box "f(u)" which calculates $2FM/(F+M)$ which is then multiplied by b_f for input to "Subtract"; and (3) to "Scope F" for its graph. The box "Female File" collects all the data so that we can present all the graphs in one figure. Chapter 28 explains how to do this. There are also other items to set/choose, like simulation time and type of numerical integrator, which we will not discuss here (see Chapter 28).

Let $\overline{F}(t)[0] = [f_1(t,0), f_2(t,0)]$. In Figure 26.2 connect the upper boundary with a smooth curve which is our approximation to $f_2(t,0)$ and connect the lower boundary with a smooth curve producing an approximation to $f_1(t,0)$. The horizontal axis is time and the vertical axis is the number of females in the population. We see that in over 1000 time units the female population can decrease or it can quickly increase. The uncertainty band after 1000 time units is extremely large at approximately $9.6(10)^8$. The three curves colored red, black and blue (see Section 1.5.4) approximately coincide in this application and show up as a single black (slightly blue in the online publication) curve.

Let $\overline{M}(t)[0] = [m_1(t,0), m_2(t,0)]$. In Figure 26.3 connect the upper boundary with a smooth curve which is our approximation to $m_2(t,0)$ and connect the lower boundary with a smooth curve producing an approximation

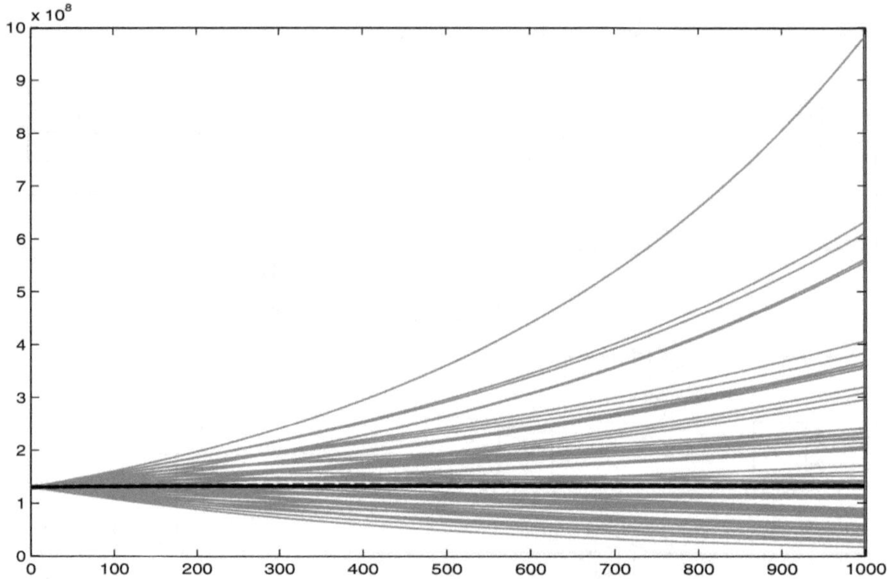

Figure 26.2: Fuzzy Trajectory for Females $\overline{F}(t)[0]$ in the Sex Structured Population Model

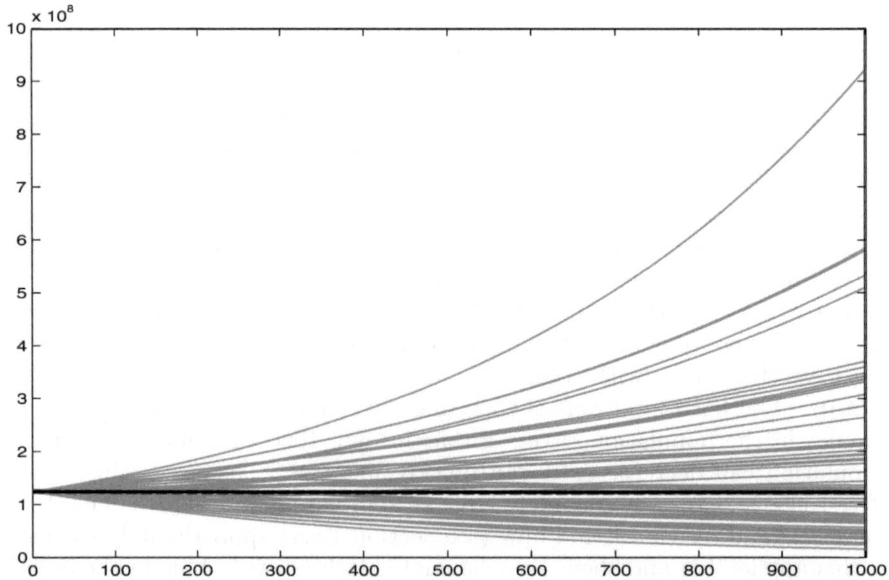

Figure 26.3: Fuzzy Trajectory for Males $\overline{M}(t)[0]$ in the Sex Structured Population Model

to $m_1(t, 0)$. The horizontal axis is time and the vertical axis is the number of males in the population. Over the interval $[0, 1000]$ the male population can continuously decrease or it can rapidly increase. The uncertainty band at $t = 1000$ is approximately $8.9(10)^8$. The three curves colored red, black and blue (see Section 1.5.4) approximately coincide in this application and show up as a single black (slightly blue in the online publication) curve.

The uncertainty band for both populations continues to grown without bound as $t \to \infty$. Since the uncertainty grows so large in increasing time it is impossible to make any realistic predictions.

If all the uncertain parameters are estimated from data (Chapter 3), so none are from expert advice, then the base of the fuzzy estimator is a 99% confidence interval. Using all of these confidence intervals we get $\overline{F}(t)[0]$ and $\overline{M}(t)[0]$, so these bands of uncertainty will be like 99% confidence intervals for $F(t)$ and $M(t)$, respectively.

26.4 Reference

1. M.Kot: Elements of Mathematical Ecology, Cambridge University Press, Cambridge, UK, 2001.

Chapter 27

Summary and Future Research

27.1 Summary

This book is about analyzing continuous time fuzzy dynamical systems (just "continuous fuzzy systems" for short). We start with a continuous time crisp (not fuzzy) dynamical system (just "continuous system" for short). This continuous system is described by a system of ordinary differential equations (ODEs). Let $x(t)$ and $y(t)$ be two variables in the system we wish to study. That is, $x(t)$ and $y(t)$ are solutions to the system of ODEs. We look at the graphs (trajectories) of $x(t)$ and $y(t)$ to see how the system evolves for increasing time. Usually the "closed form" mathematical solutions for $x(t)$ and $y(t)$ are complicated, we could have nonlinear systems of ODEs, so we will use crisp continuous simulations to obtain the graphs of $x(t)$ and $y(t)$.

The system of ODEs contains a number of parameters whose values must be known to construct the trajectories of x and y. We could know the exact values of some of these parameters but the other parameters have uncertain values. For these uncertain parameters we obtain estimates for use in the model. Usually one uses point estimates (single values) for the uncertain parameters in the system of ODEs. However, there is uncertainty in the values of these parameters which is not shown in their point estimators and the resulting crisp solutions $x(t)$ and $y(t)$. One might consider using interval estimators (possibly confidence intervals) for the uncertain parameters. We propose using fuzzy number estimators transforming the crisp system of ODEs to a fuzzy system of ODEs.

Alternatively, one could consider the values of the uncertain parameters to be generated by random variable. Then we would get a system of stochastic ODEs to solve. We used only fuzzy estimators in this book and never considered uncertain parameters to be values of random variables.

Chapter 3 discusses the two methods of fuzzy estimation we used in this book. They are: (1) from expert opinion; and (2) from crisp data. The "from crisp data" procedure involves placing confidence interval one on top of another to construct a triangular shaped fuzzy number estimator. Either method produces a triangular (shaped) fuzzy number estimator for the values of the uncertain parameters in the system. This transforms the system of ODEs into a system of fuzzy ODEs to solve. Using fuzzy parameters we get fuzzy differential equations and the solutions $\overline{x}(t)$ and $\overline{y}(t)$ are also fuzzy. The system now becomes a continuous fuzzy system whose trajectories are fuzzy so that any slice through a trajectory at some time t is a triangular shaped fuzzy number. In all the applications we use triangular fuzzy numbers, not triangular shaped fuzzy numbers, for fuzzy estimators. Triangular fuzzy number inputs makes the fuzzy trajectories triangular shaped fuzzy numbers. We wish to estimate the maximal band of uncertainty in the system which is $\overline{x}(t)[0]$ and $\overline{y}(t)[0]$, or the $\alpha = 0$ cuts. These will be the bases of the fuzzy numbers.

In the crisp system of ODEs let a_i, $i = 1, 2, 3, ..., k$, be the uncertain parameters which now have fuzzy number estimators $\overline{a}_i$, $1 \leq i \leq k$. All the other parameters are assumed to be known and crisp. Let

$$\Omega_x(t) = \{x(t) | a_i \in \overline{a}_i[0], i = 1, ..., k\}. \tag{27.1}$$

$\Omega_x(t)$ is all solutions $x(t)$ to the crisp system of ODEs for all values of the uncertain parameters in their $\alpha = 0$ cut. An objective of the book is to estimate $\Omega_x(t)$, which we also call the band of uncertainty in the solution due to the uncertainty in the values of some of the parameters.

Now $\Omega_x(t) = \overline{x}(t)[0]$. Let $\overline{x}(t)[0] = [x_1(t, 0), x_2(t, 0)]$, $t \geq 0$. The boundary of $\Omega_x(t)$ is given by $x_1(t, 0)$ (lower boundary) and $x_2(t, 0)$ (upper boundary). So we want to employ crisp continuous simulation to approximate the curves $x_i(t, 0)$, $i = 1, 2$. Consider graphing the triangular shaped fuzzy numbers $\overline{x}(t)$, $t \geq 0$, in three dimensions. For each fixed t the base of the triangular shaped fuzzy number $\overline{x}(t)$ will be on the interval $[x_1(t, 0), x_2(t, 0)]$ and its vertex at $x_1(t, 1) = x_2(t, 1)$. We do not do the graphs in three dimensions but approximate the graphs of the $x_i(t, 0)$ in two dimensions superimposing the graph of the vertex.

Finding the curves $x_i(t, 0)$ in general is an unsolved problem. We show in Section 6.2, for a second order, linear, constant coefficient ODE, the construction of $x_1(t, 0)$ and $x_2(t, 0)$ is surprisingly complicated. Therefore, we decided to employ crisp continuous simulation to obtain approximations to the boundary of $\Omega_x(t)$.

The ideal solution to approximate the boundary of $\Omega_x(t)$ would be to get the graph of $x(t)$ as the a_i go through many selected values in their $\alpha = 0$ cut, $1 \leq i \leq k$. Suppose $k = 5$ and we pick 10 values for each a_i. That gives us 10^5 graphs. Too many graphs. Each such graph takes too much time and too much space in the computer for a book with over 100 such graphs. So we had to make some compromises. This is discussed in Chapter 6. In this book the most numbers of graphs in one figure was 729 and the least was 3. We hope these figures will give the reader an indication of what $\overline{x}(t)$ looks like.

The applications are presented in Chapters 7-26. In each application we: (1) first give the system of ODEs ; (2) then discuss the fuzzy estimation of the uncertain parameters; (3) decide on how to choose the fuzzy a_i in their $\alpha = 0$ cut to approximate the boundary of $\Omega_x(t)$; (4) discuss how our continuous simulator simulated the system; and (5) present and discuss the graphs. The applications are diverse, ranging from bungee jumping to the AIDS epidemic to planetary motion. Because of the many and varied applications presented, this book provides not only an excellent introduction to the simulation of continuous fuzzy systems, but it also provides clear and usable Matlab examples on which one may build. Chapter 28 describes the Matlab/Simulink method and how one may obtain source to all the examples.

What is new in this book is: (1)we argue that many crisp continuous dynamical systems naturally become fuzzy through fuzzy estimation of the uncertain initial conditions and parameters; (2) we find an approximation to the band of uncertainty which is the trajectory of the bases of the fuzzy number trajectories; and (3) we apply this to numerous diverse applications using the readily available simulation language Simulink [2].

27.2 Future Research

Future research is to present the graphs of the fuzzy trajectories in three dimensions. But first we need to obtain better estimates of the curves $x_1(t,0)$ and $x_2(t,0)$, or the boundary of $\Omega_x(t)$.

Also, there is the possibility of a lot of future research on a number of the models we investigated in Chapters 7-26. Consider Chapter 17, the AIDS Epidemic, where there were 11 parameters (see Table 17.1) and we allowed only two of them to be fuzzy. In Chapter 17 we looked at a very limited fuzzy model. What happens if more/other parameters are fuzzy? The band of uncertainty will change but will it grow substantially?

27.3 Conclusions

This book, a companion to [1], argues that many crisp continuous time dynamical systems become fuzzy when uncertain parameters in the model have fuzzy estimators. The values of uncertain parameters must be estimated and we have shown how they can have fuzzy estimators. These two books together we hope simulates interest in fuzzy systems theory.

27.4 References

1. J.J.Buckley: Simulating Fuzzy Systems, Springer, Heidelberg, Germany, 2005.

2. www.mathworks.com

Chapter 28

Matlab/Simulink Commands for Graphs

28.1 Introduction

We chose Matlab/Simulink [1] as the tool for our simulation because of its capabilities, ease of use, availability, and cost. The diagrams presented herein were created and executed under MatLab Version 7.0.1.24704 (R14) Service Pack 1, but are believed to be compatible with earlier versions and with the student version. The diagrams are created under Simulink and will perform a simulation using the default values, but for the graphs presented, each diagram is controlled by a corresponding Matlab program.

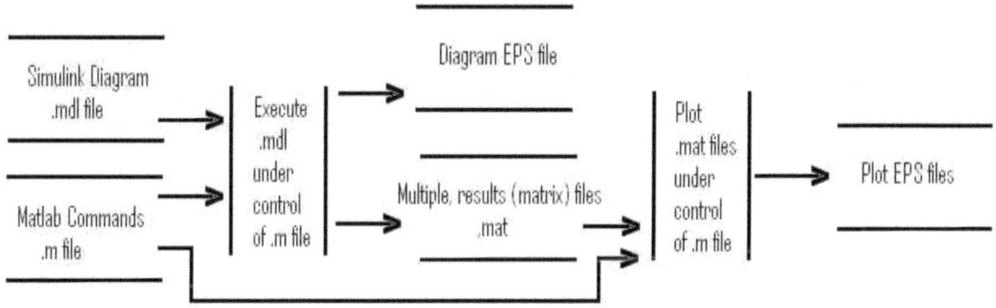

Figure 28.1: Process Flow for Simulations

As shown in Figure 28.1, the Simulink model file (.mdl) is processed by Matlab using the Matlab M-File command file (.m) to generate an encapsulated Postscript (.eps) file of the model diagram and results as Matlab matrix files (.mat). Subsequently, the same .m file processes the .mat files to generate color results plots (.eps).

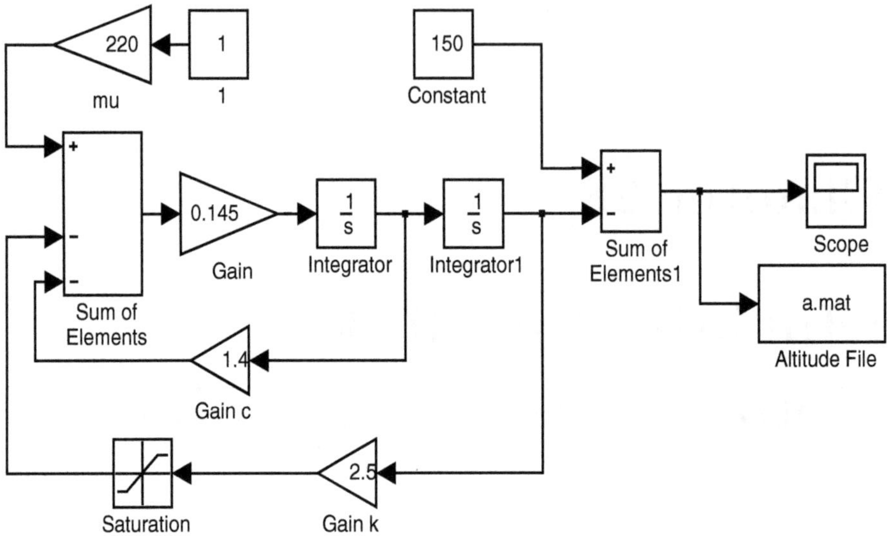

Figure 28.2: Simulink Diagram

28.2 Simulink Diagrams (.mdl files)

In our Simulink diagrams, we use basic Simulink icons and require no toolboxes. From the Matlab window, one may click on the Simulink icon to see the Library browser. Most of the library items we use are from SOURCES, SINKS, MATH OPERATIONS, CONTINUOUS (for the *Integrator* block) and DISCONTINUITIES (for the *Saturation* block). The building of the simulation diagram is straight-forward using drag-and-drop. The Bungee Jumping Example from Chapter 9, is used as an example here. Figure 28.2 is the diagram for it.

28.3 Parameters

From Matlab, many of the parameters in a Simulink diagram are modifiable (e.g., `set_param('Model.GainC','Gain',num2str(GainC));` to change 'Gain c' in Figure 28.2), but our experience was that not all are. In order to perform the multiple executions necessary for fuzzy systems, it was necessary to make *Constant* fuzzy parameters modifiable. This we did using a *Constant* 1 followed by a *Gain*; e.g., mu in Figure 28.2. The initial condition of an *Integrator* is embedded in the block as "Initial condition". We found that to modify the "Initial condition", it must be addressed as "Initialcondition". Except for these issues, control of simulation parameters was also straight-forward.

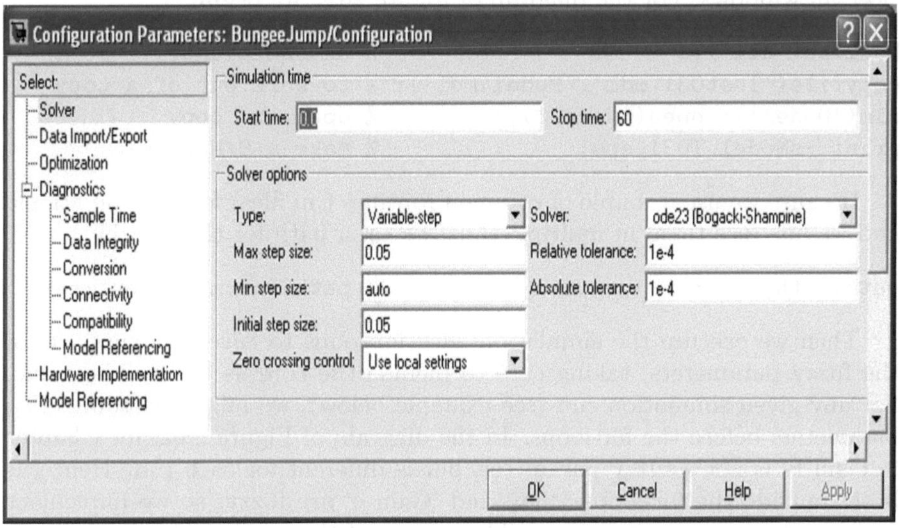

Figure 28.3: Simulink Configuration Parameters

For all of the simulation diagrams, there are some simulation configuration settings to be considered. To get to the configuration settings, from a Simulink window we click `Simulation` and then click `Configuration Parameters`. Figure 28.3 is a screen shot of the window in which the settings may be modified. After some research, tests and analysis, we chose the Bogacki-Shampine ode23 numerical integrator. Matlab provides options of solvers and describes the Bogacki-Shampine explicit continuous solver as its one of moderate complexity [1]. We found it adequate for our purposes. "Start time" starts at zero and "Stop time" is specific to each model. Rather than make the "Stop time" specific to each *run*, we sometimes "covered" the longest simulation of a set, and reported the significant period of the result. Under "Solver options," we chose our "tolerances" at 0.0001. We found a "Max step size" of 0.05 generated acceptable results without generating excessively fine plots. For "Zero crossing control", we used local settings. All other settings in configuration parameters are the Matlab default parameters. These settings we used for all simulations.

Within the "Integrator" blocks there are options also. Except to set initial values of the integrator output, we stayed with the default settings.

28.4 Matlab Commands (.m files)

The command file creates a work copy of the simulation model; this is done to protect the model from subsequent modifications. To make certain that we do not have superfluous windows or files open, we first use `bdclose` to close all windows. For the diagram of Figure 28.2, we begin:

```
bdclose('all');                        % unconditionally close
copyfile('Test031.mdl','Model.mdl');   % to work out of a copy
sim('Model'); open('Model');           % open the copy
print -sModel T031.eps;                % make a B/W eps
```

Because we use a couple of common routines (.m files) for all simulations, and we may use them in multiple runs, we set a path for them early.

```
path(path,'..\');                      % path to .m subroutines
```

Then we execute the simulation with for-loops to cover the supports of the fuzzy parameters, taking care to include the core as the middle value. For any given simulation run (see example below), we might set some crisp parameters before the for-loops. In the diagram of Figure 28.2, for Chapter 9 'Gain k' is not a fuzzy parameter, but is different for each run. Hence, it is set outside the for-loops. 'mu' and 'Gain c' are fuzzy; so we implement for-loops to cover the fuzziness. The output of the run is to be captured so the name of the Altitude file is set. Finally we execute the model.

```
filecount=1000;
set_param('Model/Gain k', 'Gain', num2str(2.5)); % ---- RUN 1
for mu=120:50:220
  set_param('Model/mu', 'Gain', num2str(mu));
  for GainC=1.0:0.2:1.4
    set_param('Model/Gain c', 'Gain', num2str(GainC));
    filecount=filecount+1;
    set_param('Model/Altitude File',
             'Filename',['data/A' num2str(filecount) '.mat'])
    sim('Model'); open('Model');
  end % Gain c
end % mu
filecount-1000              % show the number of files created.
```

The management of files could be an issue since we generate thousands of them for some simulations. We do a few things to facilitate the handling of the files. We create them in a subdirectory so they do not clutter our working directory. We assign a number (excess 1000) to each pass of the for-loops; and include that number in the name of the file(s) created.

Here we see how to invoke colorplot12 and then create a color EPS file. At the end of each simulation run, we create the appropriate result plots using one of two subroutines, colorplot12 or colorplot23. After returning from colorplot we capture a color EPS file of the plot. For most chapters we create multiple colorplots. For some, before capture, we adjust the axes using the Matlab axis command. For example, `v=axis; v(2)=12; axis(v)` would limit the x-axis to 12.

```
colorplot12('Altitude', 'data/A', filecount);
print -depsc T031r1_Altitude.eps
```

The difference between colorplot12 and colorplot23 is where in the loaded matrix the routine expects to find the data to be plotted. Both of the routines are .m files in the common area pointed to by the path statement. Because the fuzzy parameters have been emulated by symmetric triangular fuzzy numbers, colorplot can determine where the alpha=1 cut occurs for all parameters. We chose to first plot all curves in green. Then, since our parameters all start at the left support, and end at the right support, we overlay in red the curve created using left supports, in blue the curve created using right supports, and finally in black the curve created using vertices.

```
function f = colorplot12(theTitle, thePrefix, theCount)
% this colorplot is for ans(1,:) with ans(2,:)
% theTitle --- the title to put on the plot
% thePrefix --- the prefix to the contructed name of the file
% theCount --- the number of files to be used in the plot
theCount
midcount=fix(theCount/2)-500+1001
hold off; clf; %title(theTitle);
for count=1001:1:theCount
  load([thePrefix num2str(count) '.mat'])
  plot(ans(1,:),ans(2,:),'g','LineWidth',1)
  hold on;
end % count
load([thePrefix num2str(1001) '.mat'])
plot(ans(1,:),ans(2,:),'r--','LineWidth',3)
load([thePrefix num2str(theCount) '.mat'])
plot(ans(1,:),ans(2,:),'b-.','LineWidth',2)
load([thePrefix num2str(midcount) '.mat'])
plot(ans(1,:),ans(2,:),'k','LineWidth',2)
f=0;
```

Of the colorplot parameters, we do not use 'theTitle' for the book results, though titles on loose plots are very helpful in research. The first executable Matlab statement, `theCount`, because it does not terminate with a semi-colon, causes the parameter value to print in the Matlab command window; this provides a trace mechanism to monitor progress of simulations. `midcount=...` determines the index of the alpha=1 curve (recall that the count starts at 1000). To control the creation of multiple graphs, we use `hold off; clf;` to clear the last graph created, and `hold on;` later to allow the placement of additional traces on the same graph. The for-loop, sequentially for all of the captured data, loads a matrix file of curve data and plots with in green with width 1 on the same graph. Finally, the red, blue, and black curves are plotted with wider widths, for "all left supports", "all right supports", and "alpha=1" respectively. Note that the black curve may completely overlay a blue and/or red curve, and that because the width of the red is 3, some red may still be visible.

28.5 Availability of Files

All of the Matlab .mdl and .m files used to create the Simulink diagrams and result plots are available from the authors for free. They are distributed by email without other documentation or any warranty as a single 140KB zip file. For a copy of the file, please send an email request to *jowersl@cis.uab.edu*. The distribution file will be an attachment to a reply. The zip file does contain a sufficient directory structure to execute the files under Matlab/Simulink. After unzipping the distribution file, from a Matlab command window make the current directory `..\Tests`, then use the run command to execute an .m file; for example, `run 'Test012c'`.

28.6 Reference

1. http://www.mathworks.com

Index

List of Figures

List of Tables

MIX
Papier aus verantwortungsvollen Quellen
Paper from responsible sources
FSC® C105338

If you have any concerns about our products,
you can contact us on
ProductSafety@springernature.com

In case Publisher is established outside the EU,
the EU authorized representative is:
Springer Nature Customer Service Center GmbH
Europaplatz 3, 69115 Heidelberg, Germany

Printed by Libri Plureos GmbH
in Hamburg, Germany